Building, Detailing, and Converting the 1:35 Scale Italeri PT-596

By T. Garth Connelly

with input from Bob Cicconi, Alex Johnson,

George Wojcicki, Steve Sobieralski and Tracy White

Chris Daley Publishing

PT-596, shown here in a posed bow-on photo from 1945 during Squadron 40's shake-down cruise in Miami, Florida. This final wartime series of the Elco eighty-foot boats is considered the "ultimate" configuration. (Collection of T. Garth Connelly)

First published 2011

Chris Daley Publishing
365 Saint Bonaventure
Claremont, CA 91711
www.daleypublishing.com

ISBN 978-0-9841267-5-0

DEDICATION

As always, the book is dedicated to the memory of Donna Lynn "Moxie" Licsak. To paraphrase "Admiral James T. Kirk" in ***Star Trek II: The Wrath of Kahn*** … "Of my friend, I can only say this, of all the souls I've met in my travels; hers was the most … human."

It is also dedicated to my very best friend, a person who is the very embodiment of the word "friend" and who typifies the definition of the word "friendship", Nancy L. Gregoire. Nancy, if I may be permitted to quote our usual toast, "Friendship, always." Again, to use another line from ***Star Trek II: The Wrath of Kahn*** to describe Nancy's friendship "I have been and always shall be your friend."

Also, to my dear friend Martina "Marti" Montanari,

To my two friends, Kelley Hedley and her sister Amy Celli, two beautiful and caring women.

To the members of US Navy S.E.A.L Team 6, all I can say is nice job Guys! Nice job! Thank you!

T. Garth Connelly
Ellington, CT, USA
2011

TABLE OF CONTENTS

BUILDING THE *PT-596*

Italeri's plastic injection mold kit in 1:35 scale depicts *PT-596*, which was a unit of PT Squadron 40. She was nicknamed *Hell Razor* by her crew. Although shipped to the Philippines in late 1945, the squadron did not see action before the war's end. When *PT-596* came from Elco's Bayonne, New Jersey, factory, she was armed as follows:

- **One 37-mm M9 cannon**
- **One 20-mm Oerlikon gun on a Mk. XIV mount**
- **Two twin .50 caliber machine guns in Mk. XVII mounts**
- **Two Mk. 50 5-inch spin-stabilized rocket launchers**
- **Four Mk. XIII torpedoes in roll-off racks**
- **One Bofors Mk. III Bofors gun**
- **SO-3 Type radar**

It is encouraging that there is finally a "new" plastic kit available, as for the past forty years or so all that has been available has been the venerable Revell 1:72 (actually 1:71) scale release of *PT-109*. This kit has been the basis for four different kits: the *PT-109*, *PT-117, PT-167* and *PT-190*.

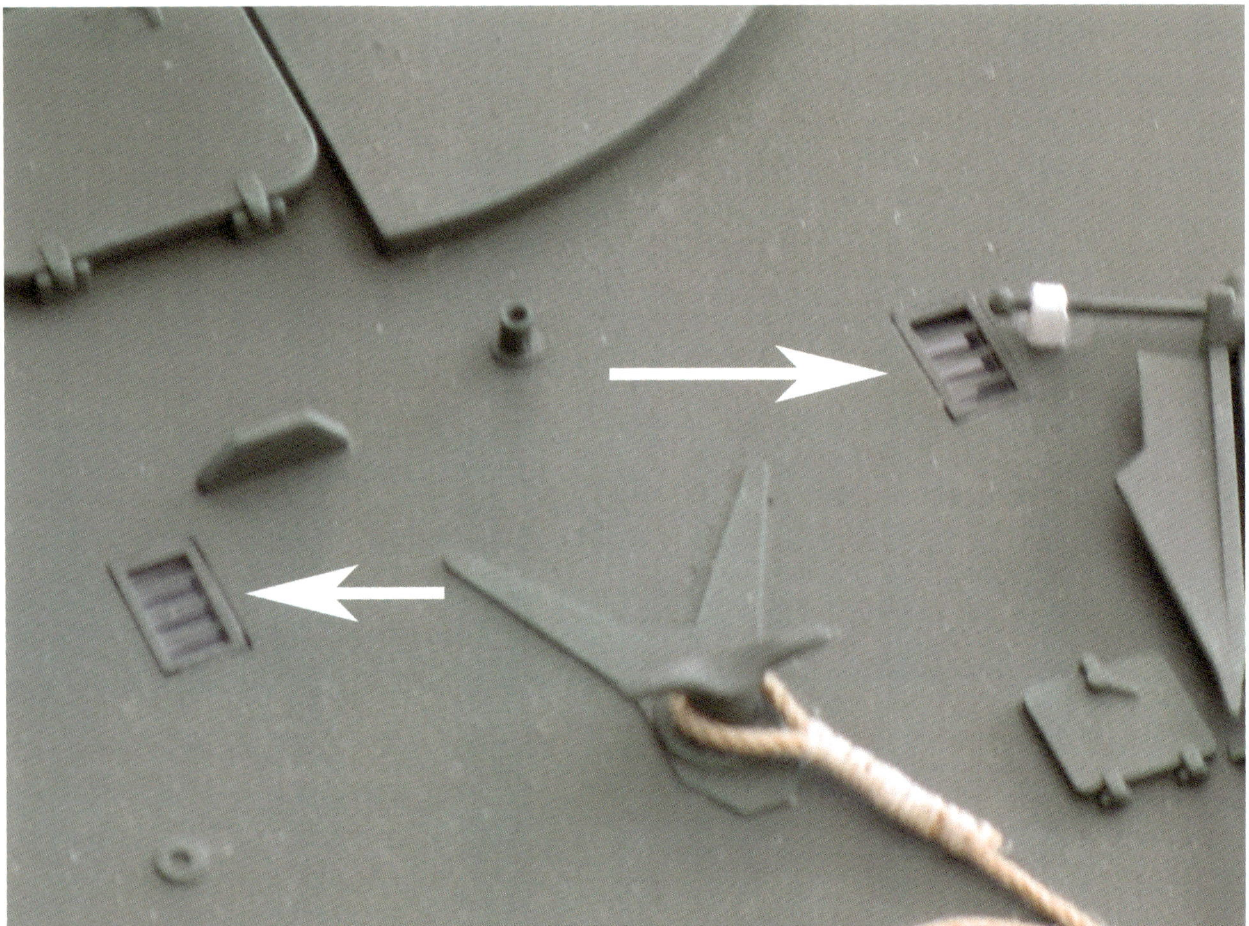

A close-up of two of the deadlights and their frames on the model's foredeck as modified. (George Wojcicki photo)

Before the *PT-109* kit, there was an earlier issue by Revell in the 1950s of the *PT-212* (in approximately 1:100 or 1:104 scale and used as the basis for the *PT-211* and *PT-207* kits). There were also the less-than-perfect plastic kits (in 1:64 and 1:32 scale) by Lindberg that were released from the 1950s through the 1990s.

That said, modelers should be aware that the current Italeri kit, impressive as it is in both size and appearance, does have some accuracy issues. For example, the color call-outs in the instructions are incorrect. Most noticeably is the calling for the life rings to be orange. The life rings should be painted to match whatever color of the camouflage scheme they were positioned in. In addition, the model in the photos was painted an incorrect shade of the base green. It should be much darker. Modelers can check a number of sources to verify the colors they wish to use for the camouflage scheme they select. Good starting places are:

- www.ptboatworld.com
- www.whiteensignmodels.com
- www.shipcamouflage.com

This view of the cockpit section of the model shows the Arming Master Switch-box which was scratch-built on the bulkhead. (George Wojcicki photo)

The scratch-built Arming Master Switch-box can be seen on the bulkhead forward of the crew figure. (George Wojcicki photo)

A minor fit issue with the Mark XIII torpedoes requires careful fitting and sanding of the racks holding them. The diameter of the torpedoes does not allow them to nestle properly in the racks otherwise. Many of the kit parts have sink-marks and/or molding marks that need to be filled and sanded before painting. In addition, almost all the tubular-shaped pieces that were molded have a noticeable seam, and many times the seams are offset.

Another error concerns the kit's decal sheet. As Bob Cicconi, who built the kit for T. Garth Connelly, noted: "The decals are not accurate regarding the size of the flag (it is too small)." He also noted that the shading on the numbers is upside-down in some places and that the kit decals had a tendency to "silver," even when put over gloss clear and using setting solutions. These errors can be corrected by after-market decals available from Archer Fine Transfers (www.archertranfers.com).

Another consulting modeler, George Wojcicki, did not like how the photo-etched brass frames for the deadlights sat on the deck. He felt they were not very flush and recommended installing the brass frames and the clear plastic pieces (which he recommends cutting to 6mm x 3.5mm) from underneath after painting the deck.

The modeler notes, in Step Three of the instructions, that there should be an "Arming Master Switch-box" on the port bulkhead of the cockpit near the instrument panel. There is a notation which states the switch should be 9mm in height. The Arming Master Switch-box was scratch-built by the modeler.

Model builder George Wojcicki put an amazing amount of detail into the cockpit area and on the instrument panel. (George Wojcicki photo)

There is a signal flag box on the front of the day-room bulkhead, just forward of and below of the dinghy and below right of the spotlight in the photo. The signal flags would have been stowed in that box when they were not in use. Mr. Wojcicki commented that there are several small dividers in the box and the flags are rolled around their shafts and stored vertically.

If the forward right window of the deckhouse were to be left open, a ladder to the roof hatch would be seen through the window. Otherwise, the modeler may want to consider depicting the boat with a blackout cover or paint the window over.

The signal flag box can be seen beneath the bow of the dinghy. (George Wojcicki photo)

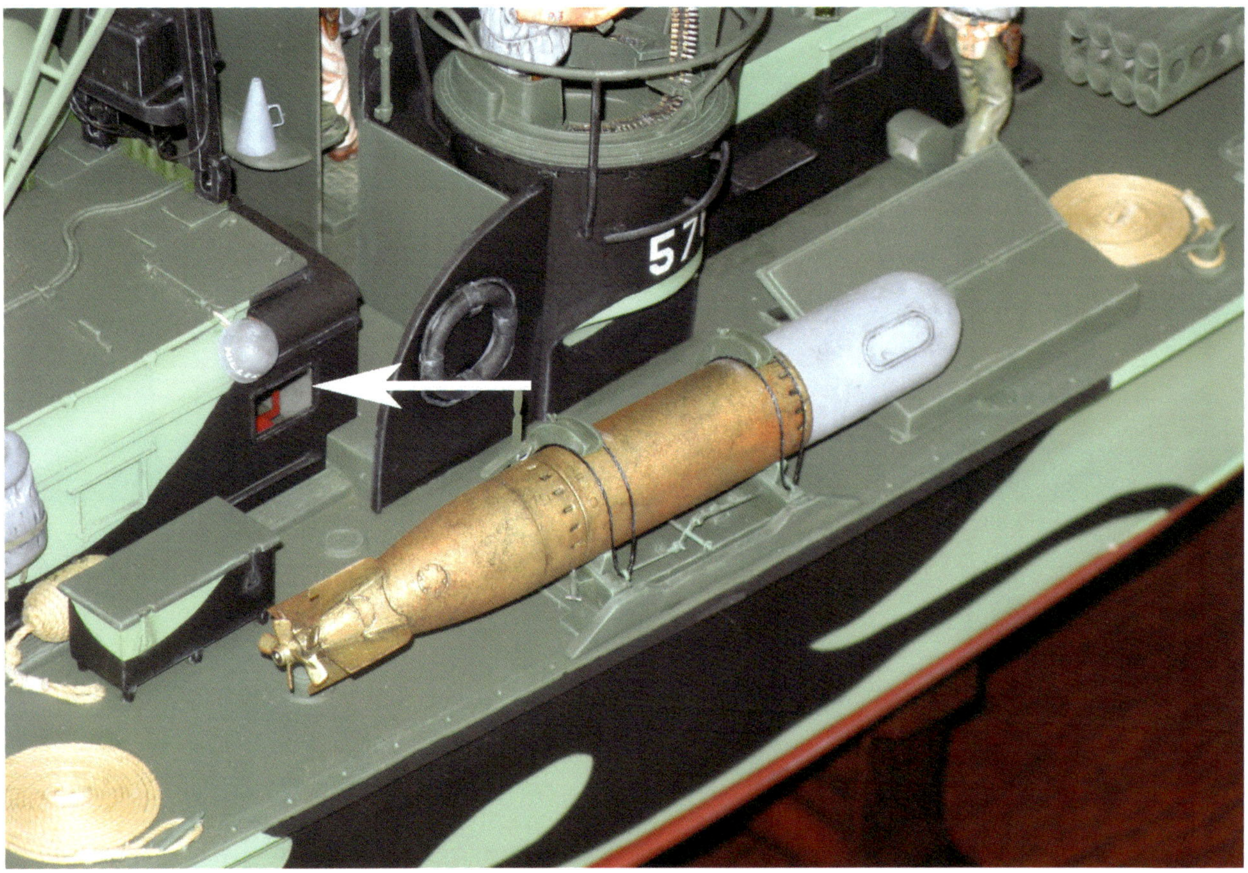

The red-painted ladder can be seen through the port in this photo just to the left of the torpedo's rudders and fins. (George Wojcicki photo)

The machine-gun tub should be closed in. Left open, the modeler can see into the cockpit. Plastic stock can be used to enclose the turret tub. A small hatch for the replenishing of ammunition is located inside the turret tub. The hatch faces into the cockpit. Looking from the rear it would be in the ten o'clock position inside the forward gun tub above the foot support.

The hatch inside the turret tub can be seen below the right elbow of the gunner. It is interesting to note that some of the shells in the belts and box are painted red to simulate tracer shells. It is the understanding of the author that every fourth shell was a tracer shell. (George Wojcicki photo)

When installing the deck structures the modeler found it useful to brace the center of the deck from the underside. This bracing will prevent the deck from bending due to its curvature during the installation of the deckhouses. During the fitting of the charthouse to the deck, the modeler had to weigh down the charthouse in order for it to seat properly. It was still necessary to fill in some gaps with putty and white glue. Depending on how much detail is desired as details on the actual boats varied as well, removal of the small mounts molded into the foredeck on the port side may be necessary. The starboard mounts do appear in period photos and were left in place. Two small cylindrical bases, which the modeler believed to be safety line poles or gun barrel rests, were added to the centerline.

In Step Seven, be sure to test fit part 39C onto the charthouse side and deck. The modeler's did not sit properly with either and required sanding on the left side and bottom. He also found that it is easier to install part 18C onto part 39C before gluing it to the deck and charthouse. Again, test fitting this step is important.

On this area of the model, a small notch has to be made on part 31D to allow for the boat hook (part 69E).

In Step Eight, the upper portions of both antennas (parts 38D, PH 7, 27D, 101E) appear lighter in color in photos and therefore were painted white and light gray.

The B.N. antenna that is located on the starboard side of the charthouse roof. (George Wojcicki photo)

As an added detail, a power cord to the searchlight was added by using thin wire. A modeler will have to use some "artistic license" as to where the cord goes. On this model, the cord ran into the cockpit along the port wall. The cord seems to connect to the light directly off the small nipple at the rear.

The searchlight on *PT-160*. (Collection of T. Garth Connelly)

The short-wave antenna can be seen on the top of the cable box. (George Wojcicki photo)

The VHF antenna can be seen behind the gunner. (George Wojcicki photo)

A close-up view of the corrected rudders. (George Wojcicki photo)

Many modelers assumed that the rudders were painted the same color as the boat's bottom. In actuality, they were left in their "natural metallic" state. (George Wojcicki photo)

The modeler discovered in drawings that there were no extensions of the propeller shafts between the strut and propeller, but since the drawings showed the propellers butting up to the struts, he installed them accordingly. Modelers will have to taper the shaft after they cut it to length to fit into the strut. The supplied rudders, parts 5E and 28C, look more like rudders from seventy-seven-foot Elco boats. To accurately represent those of the eighty-foot boats, the rudders were reshaped with a more rounded trailing edge. The rudders should be positioned closer to the bottom of the hull. Viewed from the rear the rudders should be vertical; the modeler, Mr. Wojcicki, commented that the kit forces the rudders to angle out unless one adjusts the mounting hole orientation. He found that if he did not adjust the holes, the out-board rudders wound up angled outward.

Italeri supplied some screws to assist in securing the deck to the hull. It was noted that the screws, which are supplied in the kit, are different sizes and are intended to go in different deck locations. They are also metric, so it is recommended to test fit them before trying to secure the deck.

Care should be taken to test fit the larger deck pieces as sometimes warping of the pieces was noticeable. There is a short "rod" (part 25F on the foredeck in Step Fourteen) that can be seen in period photos; however, the modeler thought Italeri's part to be too short. At first, it was thought to be an extra barrel for a 20-mm gun. However, in looking at the sprue-tree descriptions in the instructions, it seems to be one of the stanchions for the lifelines, which surrounded the 20-mm gun.

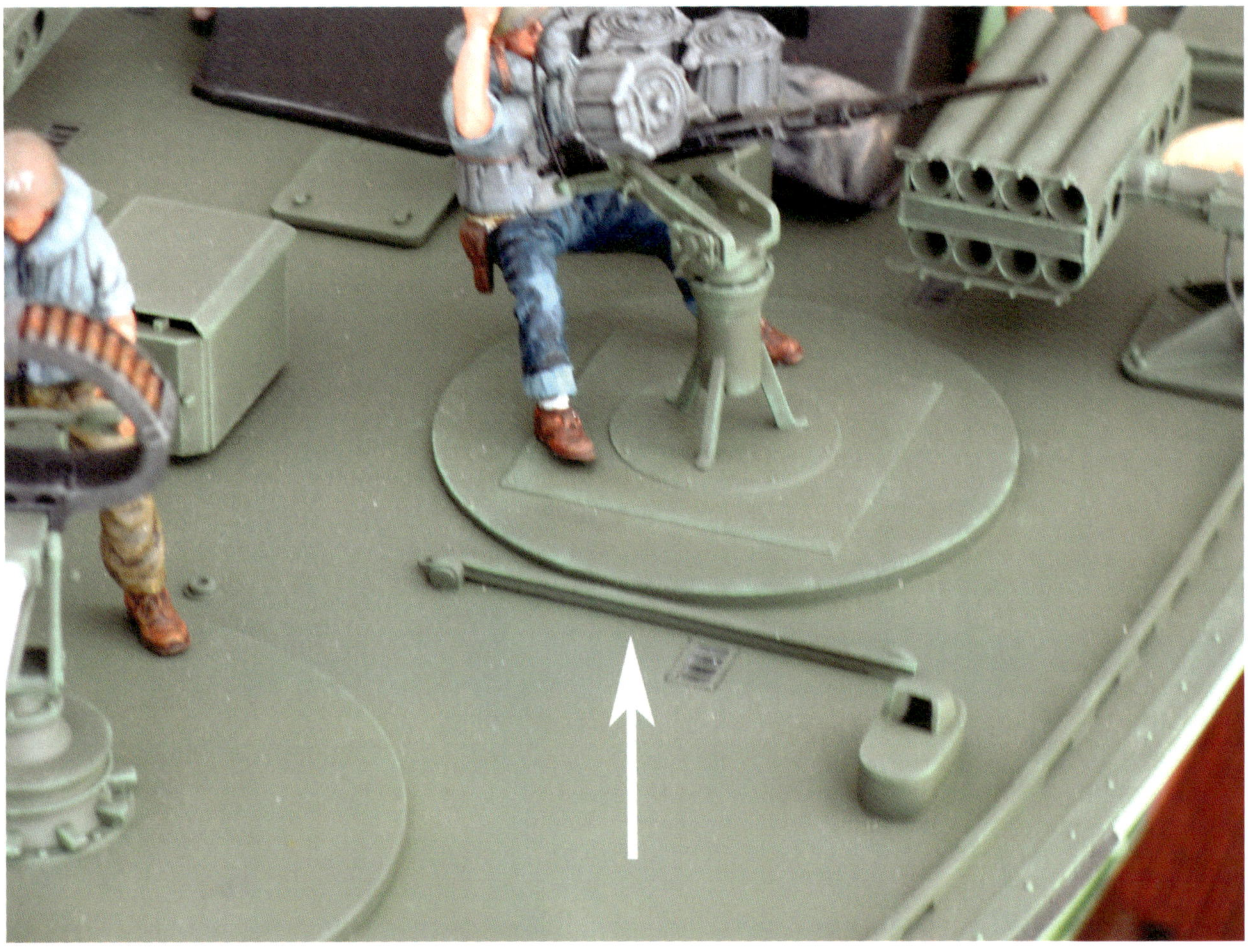

The item that was thought to be an extra barrel for the 20-mm gun is actually a stowed stanchion for a lifeline. (George Wojcicki photo)

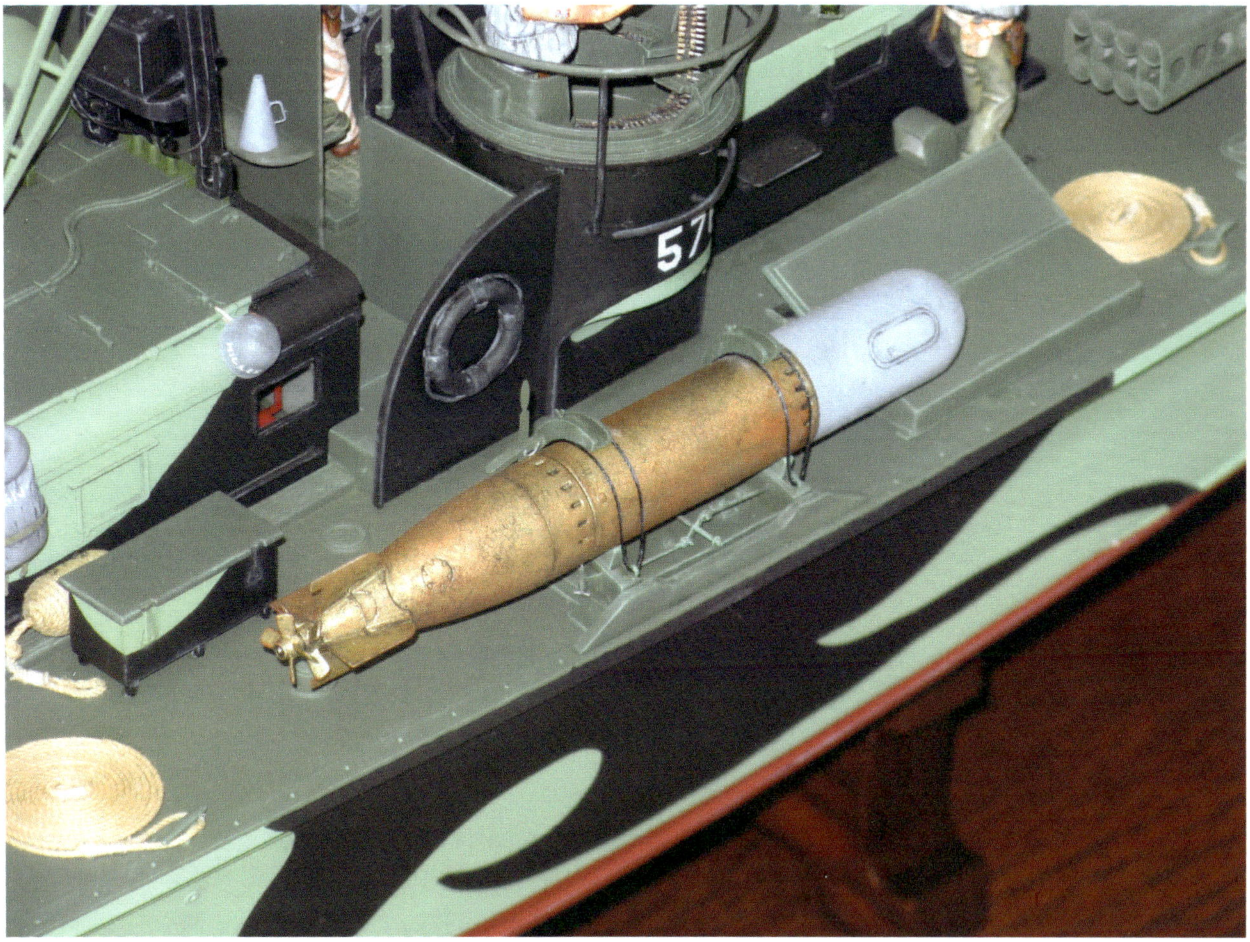

Thread was exchanged for the kit-provided wire on the torpedo racks. (George Wojcicki photo)

For the wires that held the torpedoes in places, the modeler replaced the kit's wire with black thread and found it easier to work with. When he tried to pull the supplied wires tight around the torpedoes, the small rings on the racks snapped off. A few drops of white glue on the torpedo top secured the thread before it passed through the rack rings. Care should be taken that if pulled too hard, even the thread can snap the rings. The three 20-mm magazines (parts 44E and 45E in Step Twenty) stored on one of the forward lockers should be mounted on individual brackets, which were attached to the locker top.

Three ammunition drums for the 20-mm guns were stowed atop the ready-service locker. (George Wojcicki photo)

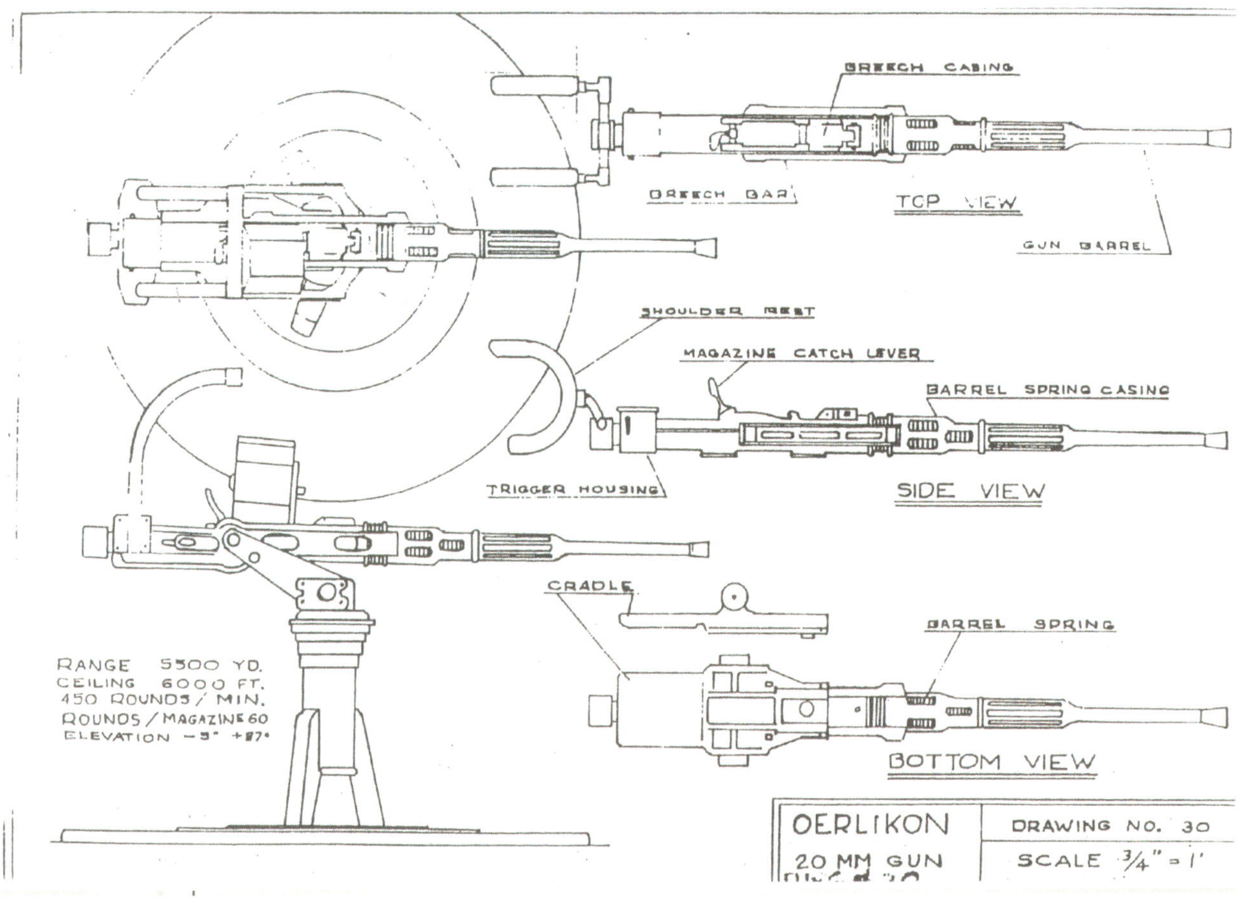

20-mm Oerlikon gun mounted on a Mark XIV mount, which was a Mark X mount with shortened legs. (PT Boats Museum and Archives, Germantown, TN)

20-mm Oerlikon gun on a Mark XIV mount on the foredeck of a late war Elco eighty-footer. (Collection of T. Garth Connelly via Frank Andruss, Sr.)

The divider/bracing in the ammunition drum for the 37-mm gun, which it was felt should be replaced, can be seen here. (George Wojcicki photo)

The modeler replaced the 37-mm magazine screen with heavier brass screen (part 20F in Step Twenty-one). It was noted that one should use a slow-setting cement when assembling the 20-mm base (part 4F) and legs (parts 72E and 19E). Install the legs on the base first and then test fit on the deck to obtain proper alignment and stance. Wartime photos show some late class boats with a tall whip antenna on top of the cable housing. This was on the portside bulkhead of the cockpit. The late war Elco boats had a thicker antenna like the one supplied with the kit.

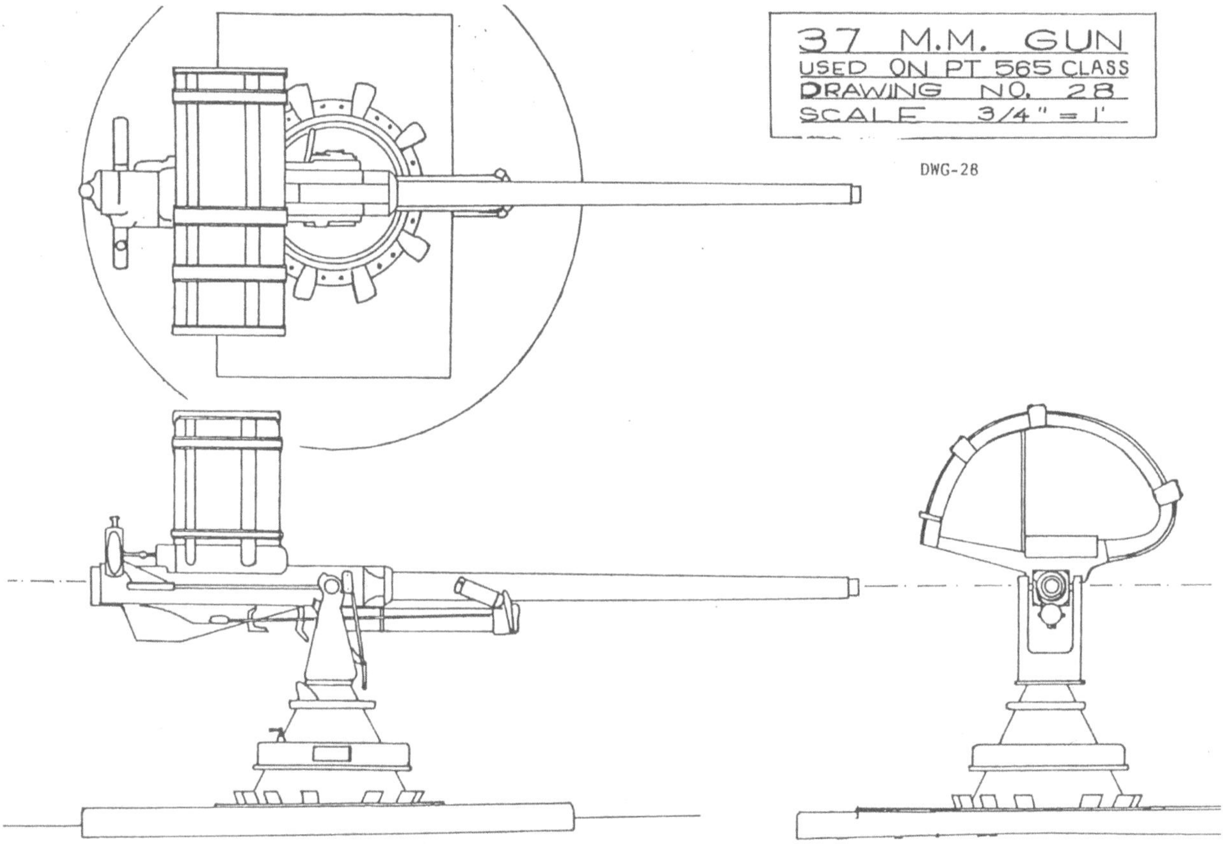

A 37-mm M9 on the 1945 Mark I factory-supplied mount, which replaced the earlier M4 version of the gun. The M4 version was fitted to simple pipe-mountings or on Mark X or XIV mounts. There is anecdotal evidence that a few earlier boats mounted a 37-mm M4 gun in place of the original 20-mm gun. (PT Boats Museum and Archives, Germantown, TN)

An Oldsmobile manufactured 37-mm M9 auto-cannon on the foredeck of an Elco eighty-footer in 1945. (Collection of T. Garth Connelly via Frank Andruss, Sr.)

This is an up-close view of the ammunition drum of a PT's 37-mm auto-cannon. Note how the ammunition drum does not have the divider, which is seen on the 37-mm gun on *PT-596*. (Alex Johnson photo)

The added detailing done by George Wojcicki can be seen to good advantage in this view. (George Wojcicki photo)

Before installing the "dinghy" and brackets (parts 73E) as shown in Step Twenty-six, it is necessary to sand and shape the rear bracket to fit the curvature of the deckhouse roof. A small "J" or "fish hook" shaped device fitted directly under the radar antenna is missing from the kit. Check the photo references for proper shape and placement; modelers may have to be a bit creative and design their own mount for it. Part 33D in Step Twenty-seven is the masthead light.

The modeler added detail to the mast. It was confusing as to why Italeri left the back of part 34D open. It can be clearly seen from behind when mounted on the mast. The plastic "ropes" were replaced with thread for a better look. The actual masts were full of wires and cables running to the electrical boxes on the mast and down to the deckhouse roof. Reference photos can be useful. Different thicknesses of copper wire were used to simulate the cables. The three "nipples" molded onto the forward deckhouse roof are power cable connectors for the mast equipment. Keep them in mind for when wiring the mast.

The SO-3 radar mast as seen from the front. The thread used to replace the molded on plastic ropes can be seen. (George Wojcicki photo)

The SO-3 radar mast as seen from the port side. The dinghy rests on the day-room roof. (George Wojcicki photo)

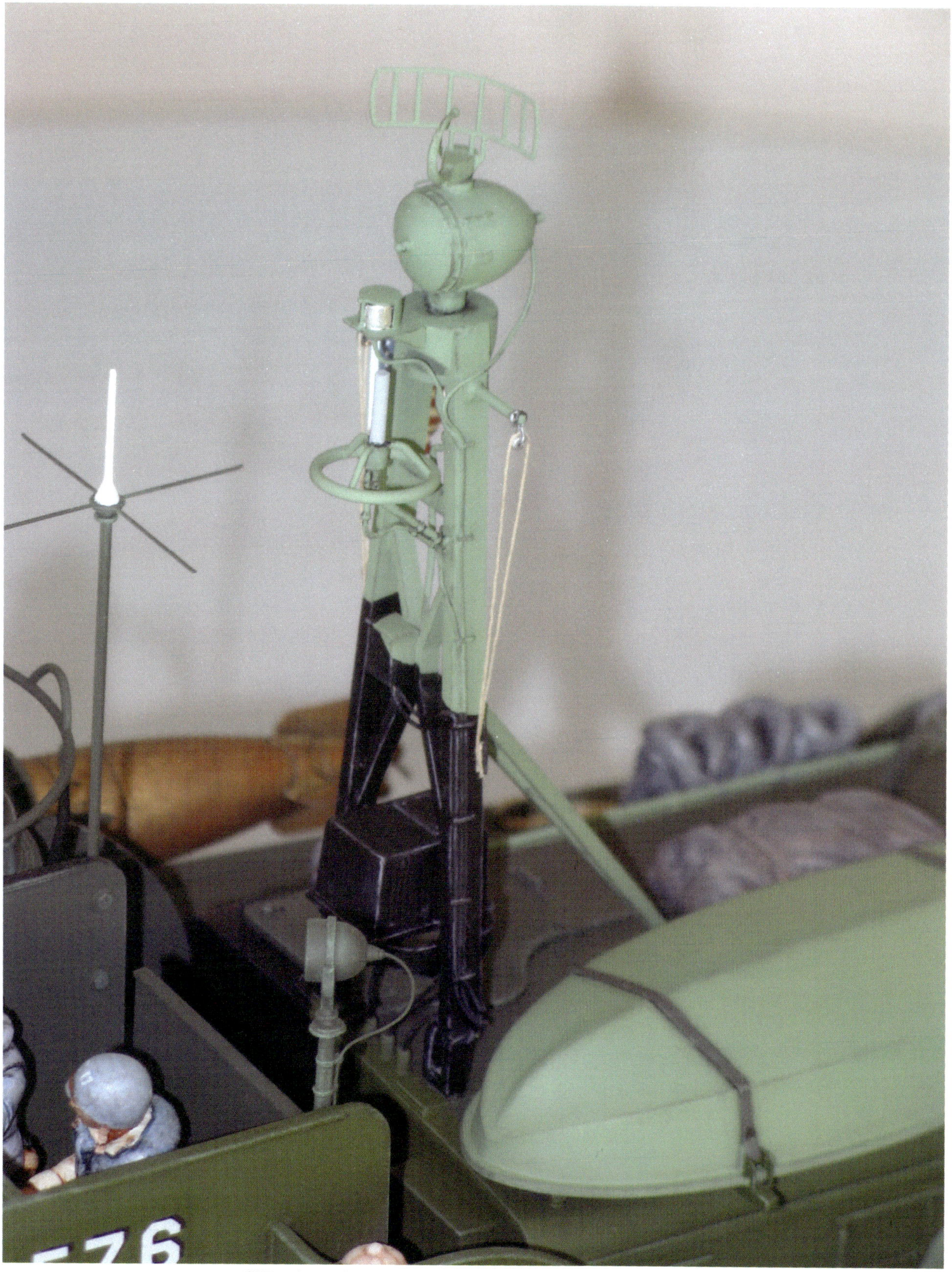

Side view of the SO-3 radar mast on the Italeri model.
(George Wojcicki photo)

Detail of the 40-mm Bofors gun on George Wojcicki's model of *PT-576*.
(George Wojcicki photo)

A close-up view of the Bofors 40-mm M3 gun crew on *PT-155*. (Courtesy of Alex Johnson)

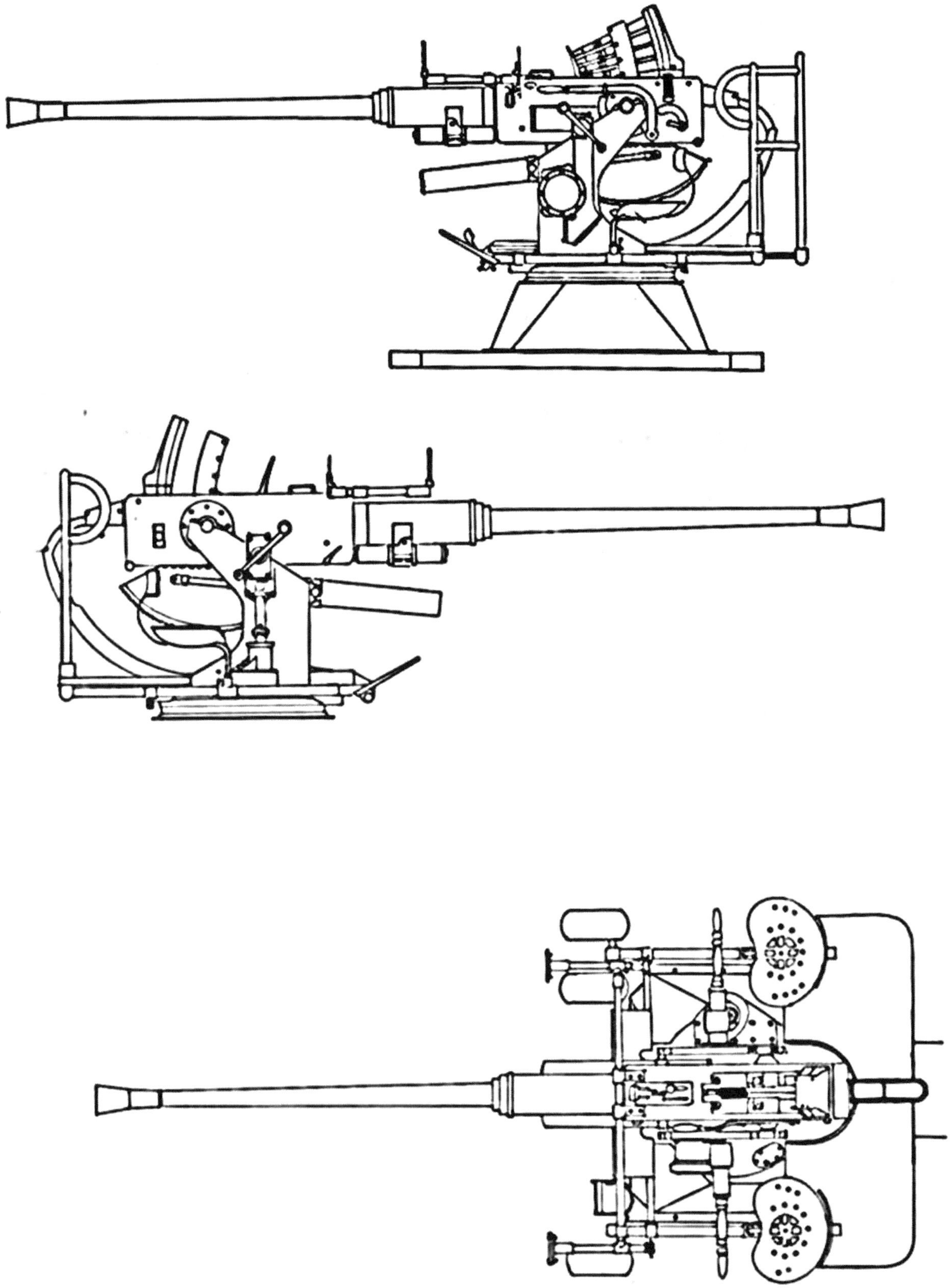

Bofors 40-mm Mark III. (Drawing by Steve Tuhy)

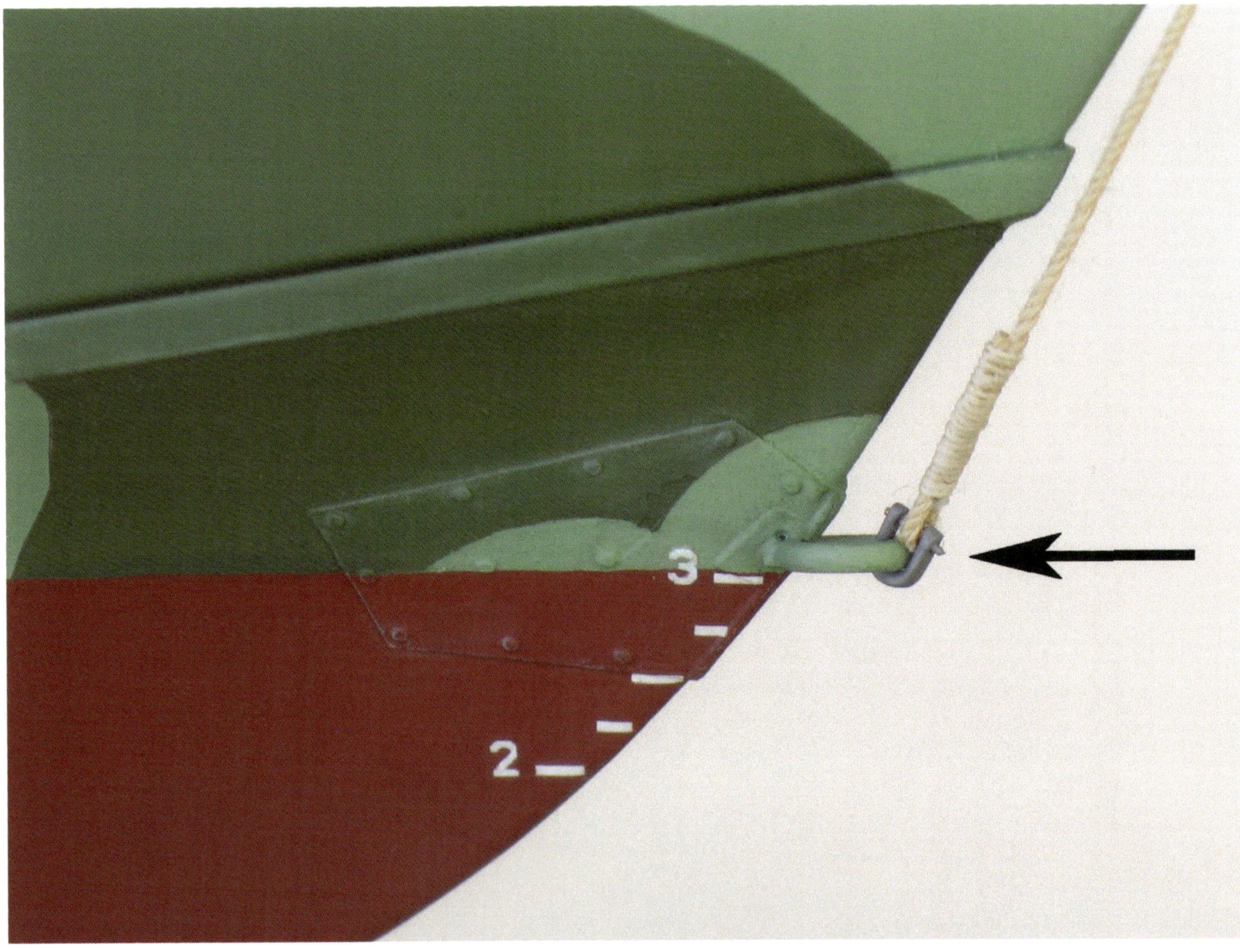

The U-shaped buckle used for the anchor line can be seen here.
(George Wojcicki photo)

Based on reference photos, the stern jack-staff had a bend in it near the base. Two small flanges are at the base of the staff, and these were attached on the forward side of the staff. There are small holes cut into each flange. When viewed from above, the whole assembly resembles a letter "Y." Between the two flanges there is a wire cable connector, which most likely is the stern light power cable.

Since the jack-staff was removable, the cable may come out the lower portion of the staff. It is noteworthy that earlier eighty-footers had this cable connection outside of the flanges on the starboard side.

Italeri incorrectly has the towing line running to the anchor. The actual line is attached to "U" shaped turnbuckles on either end. One end is secured to the bow tow-ring; the other goes through the fairlead (part 24D) and is secured to the forward most toerail cutout with a separate line.

Another of the U-shaped buckles can be seen in this photo. (George Wojcicki photo)

The anchor line was usually stowed below deck. The small hatch to the left of the Sampson post is access to the rope locker where extra rope, lines and chain were stored. The anchor was detailed by adding an additional triangular support just beyond the two flanges.

The three anchor supports appear very light in color and so were painted white. The "eye" on the anchor was angled downward so as to look as if it had weight and could swivel.

POSSIBLE CONVERSIONS FROM THIS KIT

PT-109

When the Italeri kit came out a few years ago, the kit depicted the *596* boat, which was shipped to the war zone late in the war. She never saw action against the enemy. Modelers have been wondering how they could either convert or backdate the kit into a boat which saw action in the war. Model builder Steve Sobieralski backdated the Italeri kit from *PT-596* to the series of the Elco *PT-103* Class. This series consisted of *PT-103* through to *PT-196*. The boat which he chose to depict was *PT-109* as she appeared on the day of the night she was lost in action, August 1, 1943. On that day, she had just been fitted with a 37-mm M3 anti-tank gun. Without its wheels but still on its carriage, the gun was lashed on the foredeck to two 2x8s and not lashed to coconut tree logs as it is popularly portrayed.

_PT-103_ the first of the Elco eighty-foot boats in World War II. Modelers are faced with a massive amount of work in order to backdate the Italeri kit to this series of boat, but it can be done successfully. (Collection of T. Garth Connelly via Frank Andruss, Sr.)

Over-head view of Steve Sobieralski's *PT-109*. The mast is depicted as folded down. There has been debate over whether or not the *109* even had a mast. Several photos show no mast. It is the opinion of the author that the *109* did have a mast but that it was folded down in those photos. (Steve Sobieralski photo)

It was necessary to rework the cockpit section as the "design" had been modified slightly from the *PT-103* series to the *PT-565* series. A needed modification would be to move the machine gun turret forward by three scale feet. Remove the forward leaning angle along with the indentation in the turret.

As built from the box, the Italeri kit depicts *PT-596*. *PT-109*, on the other hand, was one of the earliest eighty-foot boats, and though the early and late boats share the same hull, running gear and basic layout above the deck line, there are actually many detail differences between the two.

Apart from the hull, propellers, shafts, rudders, and mufflers, the only parts from the kit that could be used unmodified were the smoke generator, .50 caliber machine guns and mounts, and some deck fittings such as cleats and bitts. Everything else had to be either modified from the kit parts or built from scratch.

Detailed photo of an early Elco eighty-footer's charthouse and cockpit section. (Collection of T. Garth Connelly)

Looking aft at the charthouse of *PT-107*. (Collection of T. Garth Connelly)

A view of *PT-121*'s charthouse and cockpit while in New Guinea. The charthouse and cockpit area differ from the *107*. Note that the mast either was folded down or was removed. (Collection of T. Garth Connelly)

Detail of the cockpit section and the changes made. (Steve Sobieralski photo)

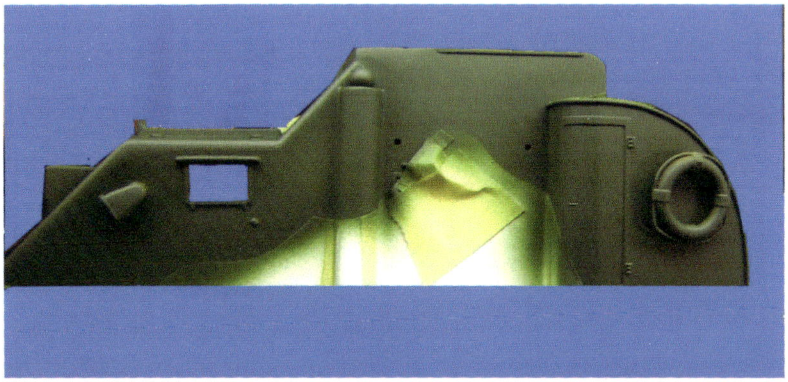

The cockpit section from the kit as built by Bob Cicconi clearly shows the differences between the earlier boats and the one of a late war boat. (Bob Cicconi photo)

The first step in the conversion was to remove most of the molded-on surface detail from the bridge/charthouse, day cabin and engine room hatch. Most of this detail is not correct for the earlier boats. Modifications to the bridge/charthouse included moving the .50 caliber machine gun turret forward, filling in the cut-away on the lower portion of the turret, and reworking the spray shields around the bridge and at the rear of the structure. Also changed was the radio antenna base configuration from round to rectangular. The four torpedo tubes required a major scratch-building effort as each one contains over 200 individual pieces. The modeler used 3/8" CPVC plumbing pipe, which is almost the correct in diameter, for the tubes. Details were added using Evergreen plastic sheet, strips and rods. He was able to expedite the process somewhat by making a master of the breech, which has some rather complex webbed bracing, and casting copies in resin. The photo below shows this work in progress as well as the foundations for the torpedo tubes and the beginnings of the early style mast.

The day-room on Steve Sobieralski's *PT-109* model. (Steve Sobieralski photo)

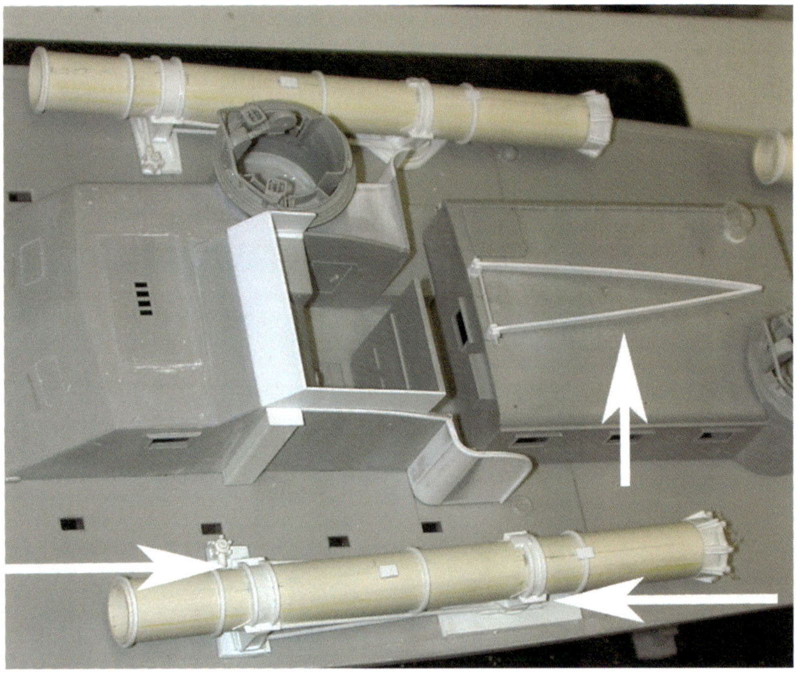

The foundations for the torpedo tubes, along with the beginnings of the early style mast. (Steve Sobieralski photo)

At the bow, additional details have been added to the charthouse roof, including railings, roof hatch and horn. Deck details, including scratch-built foredeck ventilators and deadlight frames with two instead of the three dividers provided in the kit, have also been added. Painting has commenced on some of the sub-assemblies, and areas of the charthouse have been pre-shaded prior to painting. At the stern, various additions and modifications to the kit have been made. A window and spray shields have been added to the rear of the day cabin, and coamings have been added to the sides of the engine room hatch. An early-style engine room companionway has been built to replace the large combination ventilator/hatch on the later boats.

The kit provides two tall cowl vents, but the earlier boats had four. A kit part was used to make a mold and cast two more for the engine room hatch as well as a shorter one for the day cabin roof. The 20-mm cannon was used from the kit, with some scratch-built enhancements for the sight and shoulder rests. An early-style conical base was also fabricated, as well as the brackets and tubing for the gun rail. The smoke generator provided in the kit was used as-is but relocated to the center of the stern with some photo-etched brass details added from the Eduard *PT-596* set.

Modifications made to the Italeri kit to achieve a model of *PT-109*. (Steve Sobieralski photo)

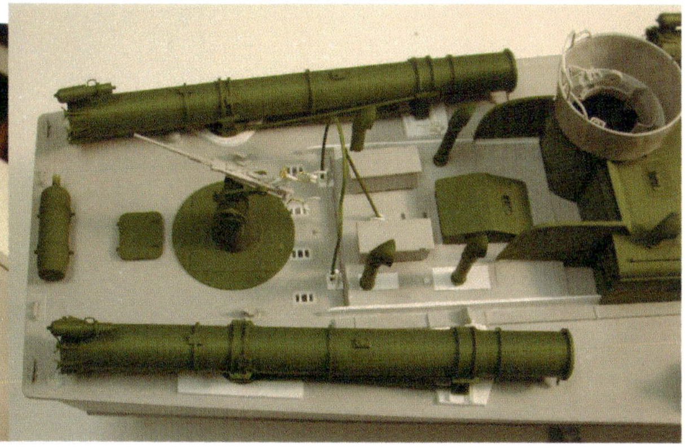

Modifications to the aft section and the 20-mm Oerlikon gun. (Steve Sobieralski photo)

The Eduard photo-etched brass set for *PT-596* was used along with their 1/35 .50 caliber ammunition belts for the gun turrets. Unfortunately, the photo-etched brass deadlight frames provided with the kit could not be used, as they were the later three-bar type. It was necessary to scratch-build the earlier two-bar type from plastic strips. The 1/35 resin kit by Trakz of a 37-mm M3 anti-tank gun is an excellent addition.

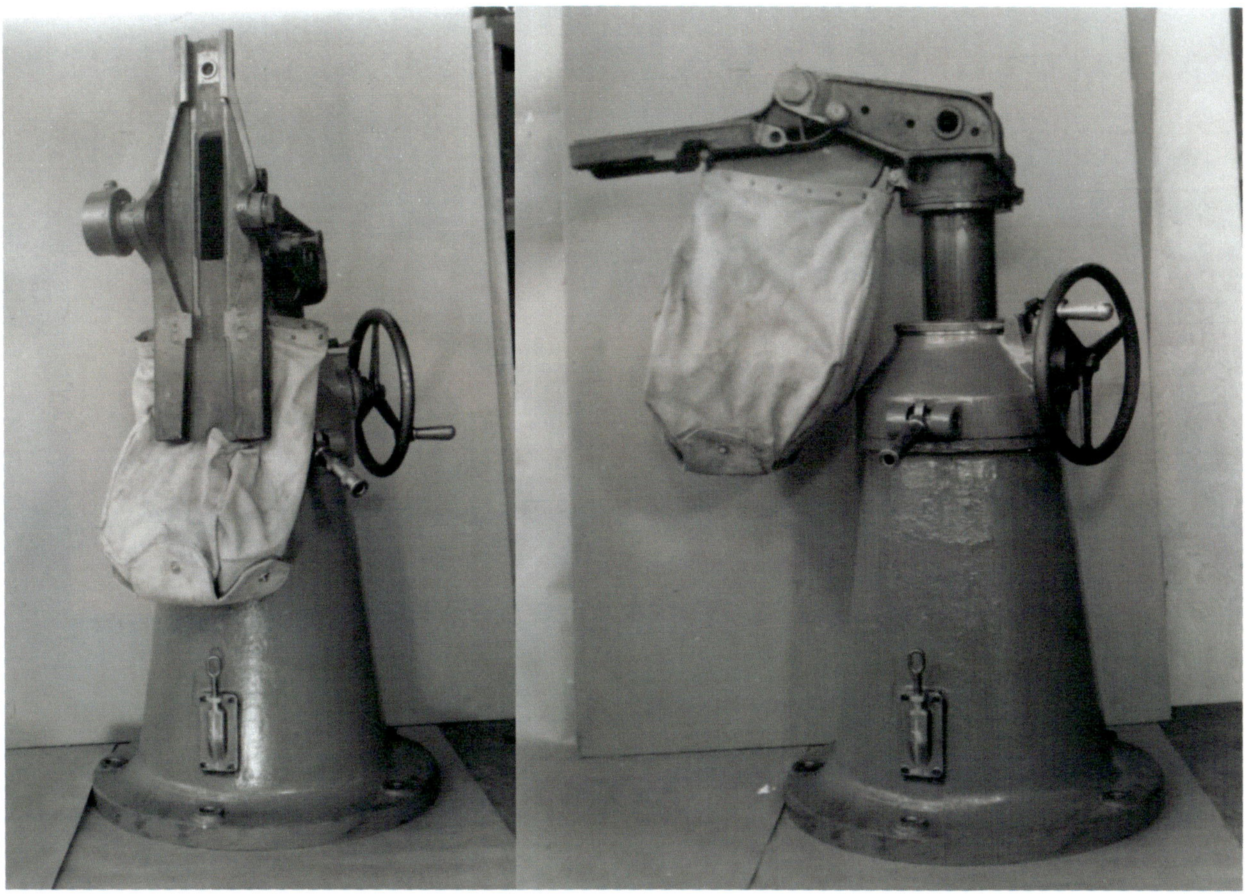

Mark IV mount for the 20-mm guns fitted to early PT boats. Some Elco eighty-footers had 37-mm M4 guns mounted on this mount for a short time. (Collection of T. Garth Connelly)

A 20-mm Oerlikon gun as mounted on a Mark IV mount. (Collection of T. Garth Connelly)

The 37-mm M3 cannon from Trakz as used on the *109*. (Steve Sobieralski photo)

View of the aft section. (Steve Sobieralski photo)

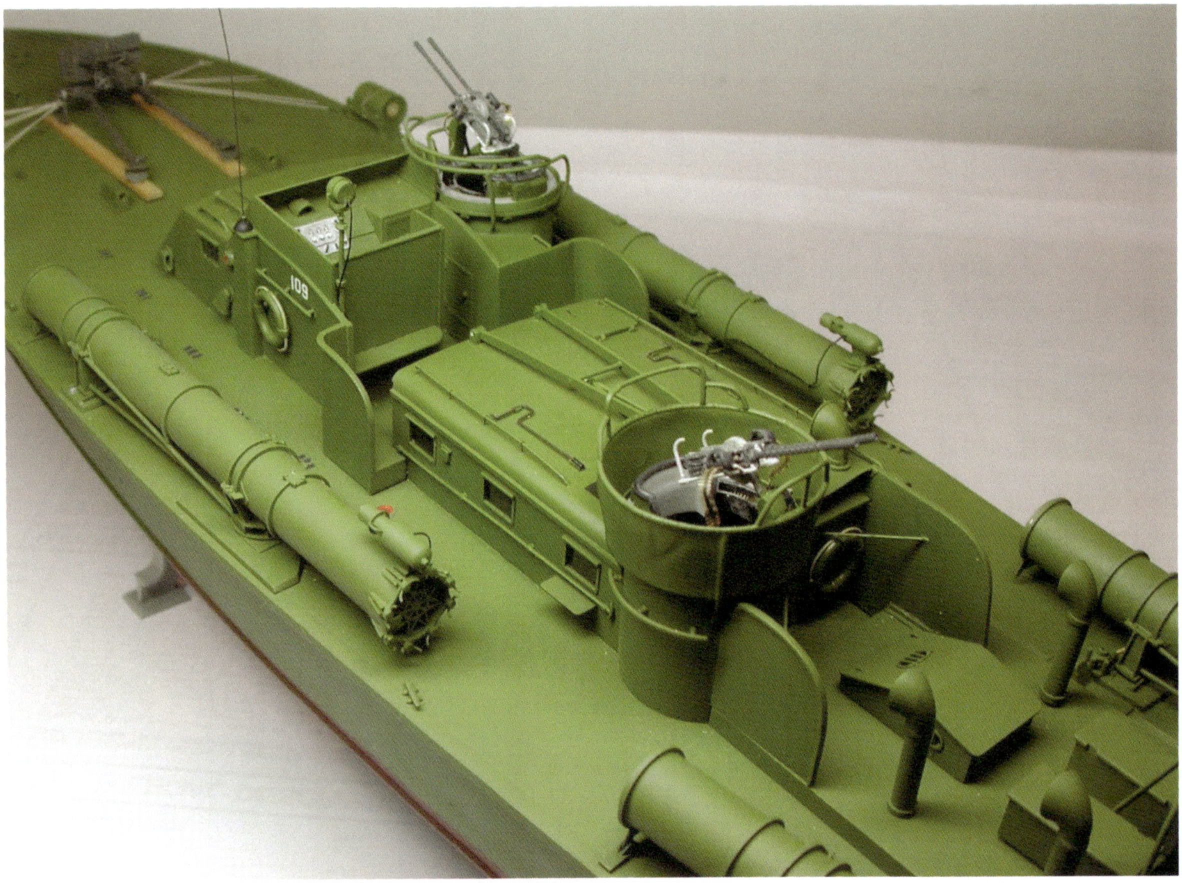

An excellent depiction of the amidships section of the model. (Steve Sobieralski photo)

Building a truly accurate replica of *PT-109* on the night she was sunk is not possible due to the lack of documentation. This model is at least consistent with the few photos and descriptions of *PT-109* that are available.

There are just too many unknowns and too few photographs or eyewitness accounts of exactly what she may have looked like. As one of dozens of PT boats operating in the South Pacific at that time, there was nothing particularly unique or interesting about her. The main references used for the basic *PT-103* type boat were the plans and manual from the 1/48 Bluejacket kit, *Allied Coastal Forces, Volume II*, along with other well-known books and publications on the subject.

PT- 559

Note: The photos of *PT-559* and *PT-557* in this section show them in the configurations they were in while serving as training boats with PT Squadron 4 at Melville. The photos seen here were taken in 1945. During this time, their weapon suites were augmented with weapons such as 37-mm guns, Mark 50 rocket launchers and depth charges. These weapons were not fitted to the boats when they had been units of PT Squadron 29.

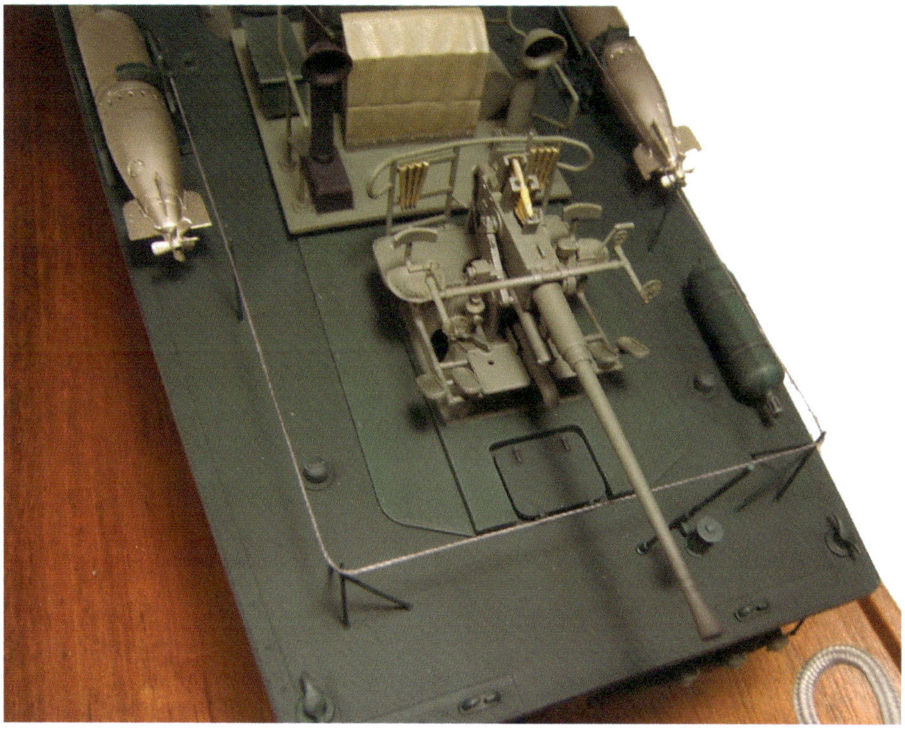

The kit provided 40-mm gun as used by Bob Cicconi, illustrates the difference of the aft gun placement of an early eighty-footer [left]] and a late war boat. (Bob Cicconi photo)

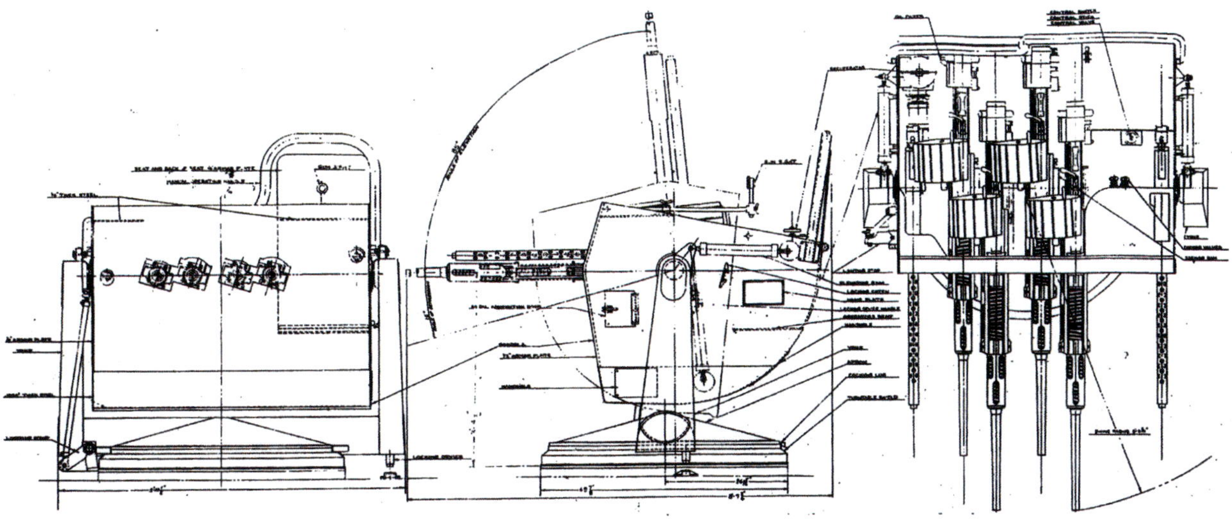

Mark XV quad-20-mm mount as fitted to PTs did not have the two .50 caliber Browning M2 machine guns. Rather, they had four 20-mm guns. (Drawing used with permission from Wayne Traxel)

Mark XV quad-20-mm mount known as the Elco Thunderbolt. (Collection of T. Garth Connelly)

There are a number of other conversions which a modeler could perform to the Italeri kit. One is the *PT-559* of Squadron 29. Steve Sobieralski also built the Italeri kit as the *PT-559*, which had an unusual weapon fitted to it. This weapon known as the quad-20-mm Mark XV mount was installed on four of Squadron 29's boats, the *556*, the *557*, the *558* and the *559*. It was called the Elco Thunderbolt.

***PT-556* seen here in Measure 31/5P before being repainted in the Thayer Blue System. (Wayne Traxell photo)**

In addition to backdating the Italeri kit into *PT-109*, the modeler also converted the kit into *PT-559* and did a masterful job. This conversion was based on an observation by modeler/historian Wayne Traxel that the basic structural layout of this kit lends itself beautifully to the boats of Squadron 27. *PT-372* was the first boat to incorporate structural features found in the *PT-596* and all later Elco squadrons in the Pacific and Europe that saw combat.

Therefore, if a modeler would want to backdate the Italeri kit to one of the boats, which were fitted with the quad-20-mm Mark XV mount, it would be quite possible to do so.

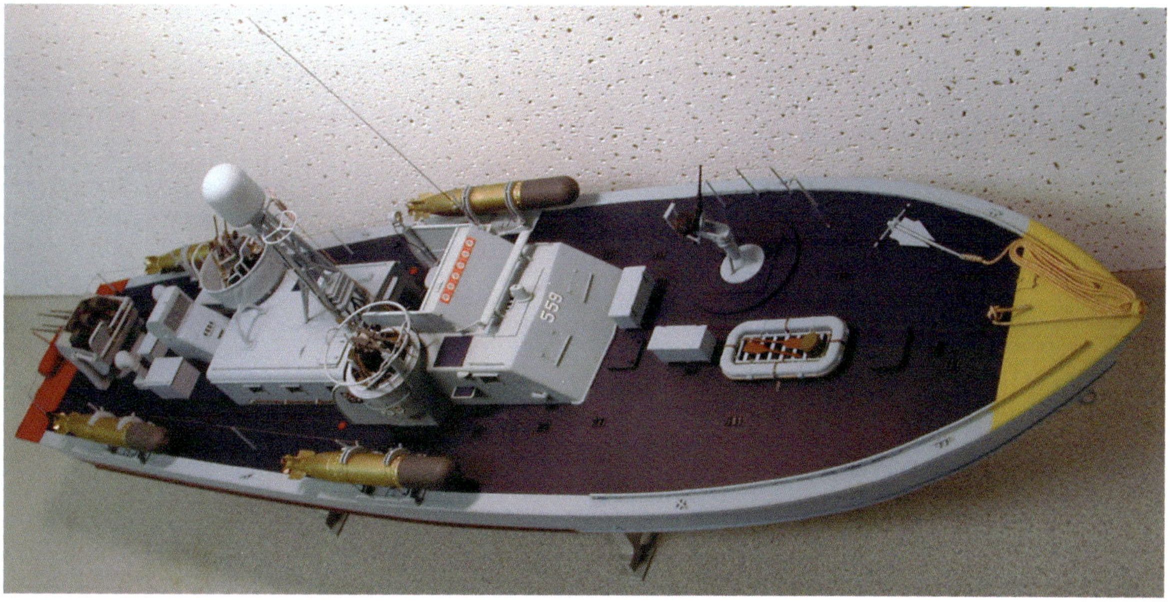

Overhead view of an excellent conversion of the Italeri *PT-596* kit to *PT-559*. The yellow triangle-shaped panel was for recognition purposes from the air. (Steve Sobieralski photo)

Two of PT Squadron 29's boats moored next to each other. The planking of the decks can be seen on each boat. It appears that neither boat's decks appear to be painted in Deck Blue (20-B). Note that the boat has the squadron's insignia on the port "bridge-wing" whereas the *559* does not have it on the starboard one. (Jerry Gilmartin)

This model is a conversion of the 1/35 Italeri *PT-596* to *PT-559*, as she was when stationed in the Mediterranean with PT Squadron 29 at Bastia, Corsica. One attractive aspect of these boats is the Thayer Blue/Deck Blue color scheme, which is a welcome change from the jungle green schemes of the Pacific Theater PT. Although the *559* was an earlier series than the *596*, the fact that it could be built using many of the kit parts was a big plus. The modeler's original intent was to build *PT-560* as one of the 40-mm equipped Squadron 29 boats. He was looking forward to using the 40-mm gun that the Italeri kit provides. He soon realized that *PT-559* was the better-documented boat in the references to which he had access. The difficulty was that this was one of four boats equipped with the power-operated quad-20-mm Thunderbolt mounting, and that would require building a scratch-built example.

Modeler Sobieralski had read that the Bastia based boats wore a four foot yellow stripe on the foredeck and a four foot wide red stripe at the stern for air recognition. The yellow painted foredeck was well documented, but not the red stern. The only photo of a boat with a discernable painted stern that was found was one of *PT-559* leaving Bastia harbor in Frank Johnson's *United States PT Boats of World War II*. There is also a close-up photo of the *559*'s charthouse and bridge area in Johnson's book, which showed several interesting details. Finding these photos clinched Mr. Sobieralski's decision to build the *559*.

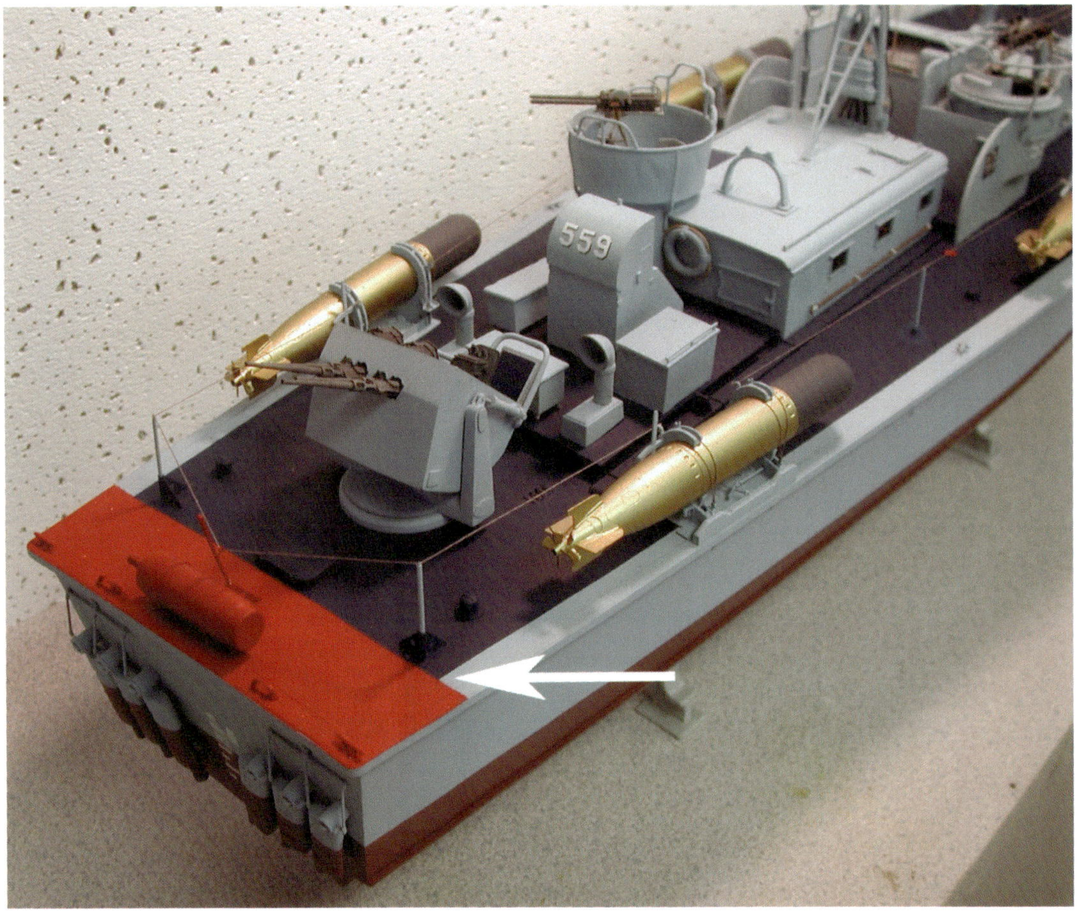

The aft section with the scratch-built Elco Thunderbolt Mark XV quad-20-mm gun mount. Photographic evidence does now indicate that some boats in the Mediterranean did have a red stripe across the aft area of the deck for air-recognition. (Steve Sobieralski photo)

Mr. Sobieralski's next problem was finding enough information to build a Thunderbolt mounting. Along with photos in various books, he found some drawings on www.ptboatworld.com that he was able to reproduce in 1/35 scale to obtain the basic shapes and sizes of the base and shield. The photos below show the shield, cradle and base. All were constructed using sheet plastic and tubing. The shield was made from .015" material, and the openings for the four guns were cut in the front plate. Two rounded rectangular holes were cut in the back, and the guns were mounted in pairs and at a slight angle to the horizontal.

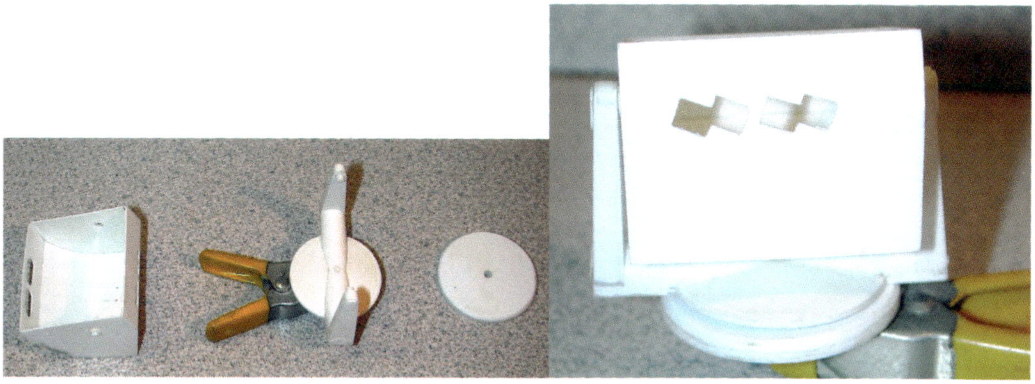

Two views showing how the scratch-built Thunderbolt mount was made. (Steve Sobieralski photo)

The torpedo racks from the kit were used with some Eduard photo-etched details. The only change made to the torpedo layout from the *596* is that the fore racks are mounted slightly further forward and the aft racks slightly further aft. This was done to allow use of the earlier Mark VIII torpedoes, which were longer, as well as the Mark XIIIs commonly associated with the roll-off racks. The engine room ventilator/hatch was narrower on the Squadron 29 boats than on the later *PT-565* series, but it was the same shape in profile.

PT-559 **in September 1944 showing her Thunderbolt mount on the stern. She is wearing Measure 31/5P. (Collection of T. Garth Connelly)**

It was a simple matter to trim the excess from the front and back pieces of the vent and attach the sides. There is a large rectangular heater box located on the starboard side of the deck hatch. Photos of the *559* taken after the war show that she had this box, but the modeler could not find any details other than that there was a grab-rail mounted outboard on the top. Work on the Thunderbolt included the gunner's seat, hydraulic elevating rams, trays for the guns, and handrail at the back. These could be used by crew members to manually elevate and train the mount.

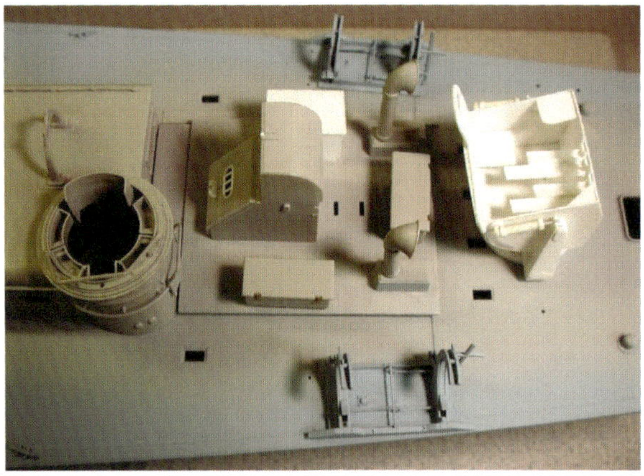

Detail of the model's aft section under construction. (Steve Sobieralski photo)

PT-557 in 1945, looking forward from the "Thunderbolt" mount. The SO-Type radar mast is folded and the aft turret does not have a normally seen canvas dodger. (Collection of T. Garth Connelly via Frank Andruss, Sr.)

The photo below shows the modifications made to the charthouse/bridge area. These included an adjustable airfoil that was fitted as an air deflector in lieu of a windshield. This arrangement was fitted to the *PT-486* through to *PT-563* series. When not in use, the airfoil was stowed on brackets atop the charthouse roof. A good deal of surface detail had to be removed or modified, such as the raft brackets and the rectangular deadlight on top of the charthouse. The bridge had to be converted to the earlier style with the throttles on the port side of the wheel and the older style instrument panel.

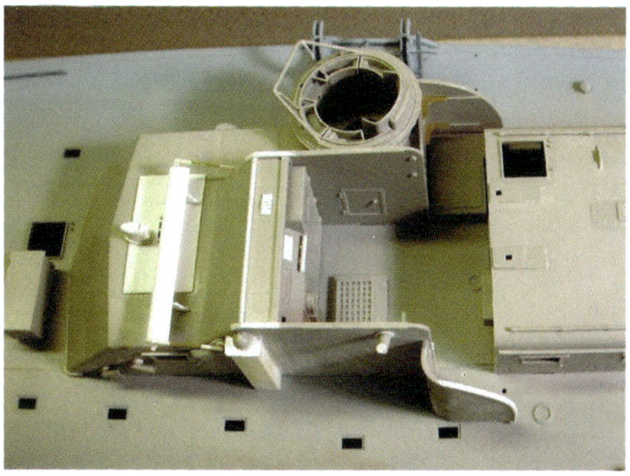

Modifications to the cockpit section can be seen here. The airfoil arrangement could be raised to deflect the wind. (Steve Sobieralski photo)

A new charthouse roof with oval deadlight, rectangular antenna base, mounting base for the torpedo sight, grab-rail for the cockpit, and bracket for the searchlight on the cockpit spray shield were all scratch-built. The streamlined shape on the roof of the charthouse is a binnacle, which housed a compass. The compass could be used by the helmsman while piloting the boat and could also be seen by the navigator through the deadlight on the roof.

***PT-507* saw action in the English Channel with Squadron 34 but was not in the same squadron as *PT-559*. The airfoil-shaped wind deflector was employed on numerous boats. (Collection of T. Garth Connelly)**

Photo taken aboard *PT-507* showing the wind deflector from a different angle. (Collection of T. Garth Connelly)

The stern of *PT-557* showing a Thunderbolt, depth charge racks, and mufflers. (Collection of T. Garth Connelly via Frank Andruss, Sr.)

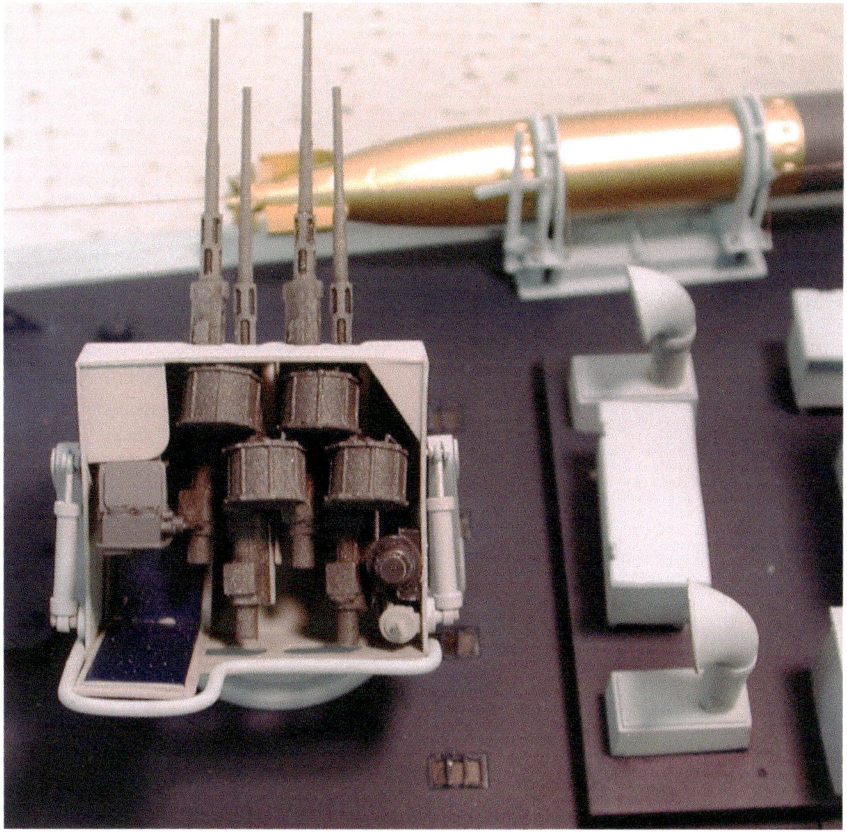

(Steve Sobieralski photo)

The above photo shows the completed Thunderbolt mount. The modeler searched the internet and was able to find photos of a model of the Thunderbolt, built at the Elco shop during the war. In the end, some guesswork was required, but fortunately, the four 20-mm cannons conceal most of the interior of the mount. A mold was made of the 20-mm gun from the kit and used to cast four more in resin. Since the barrels did not cast as well as desired, the modeler found some useful machined metal 1/35 20-mm barrels on eBay. The rest of the mount and sight was fabricated from Evergreen plastic sheet and shapes.

***PT-174* wearing Measure 31/20L in the Philippines in 1945 armed with the quad-20-mm Thunderbolt mount. (From www.ptboatworld.com, courtesy of Wayne Traxel)**

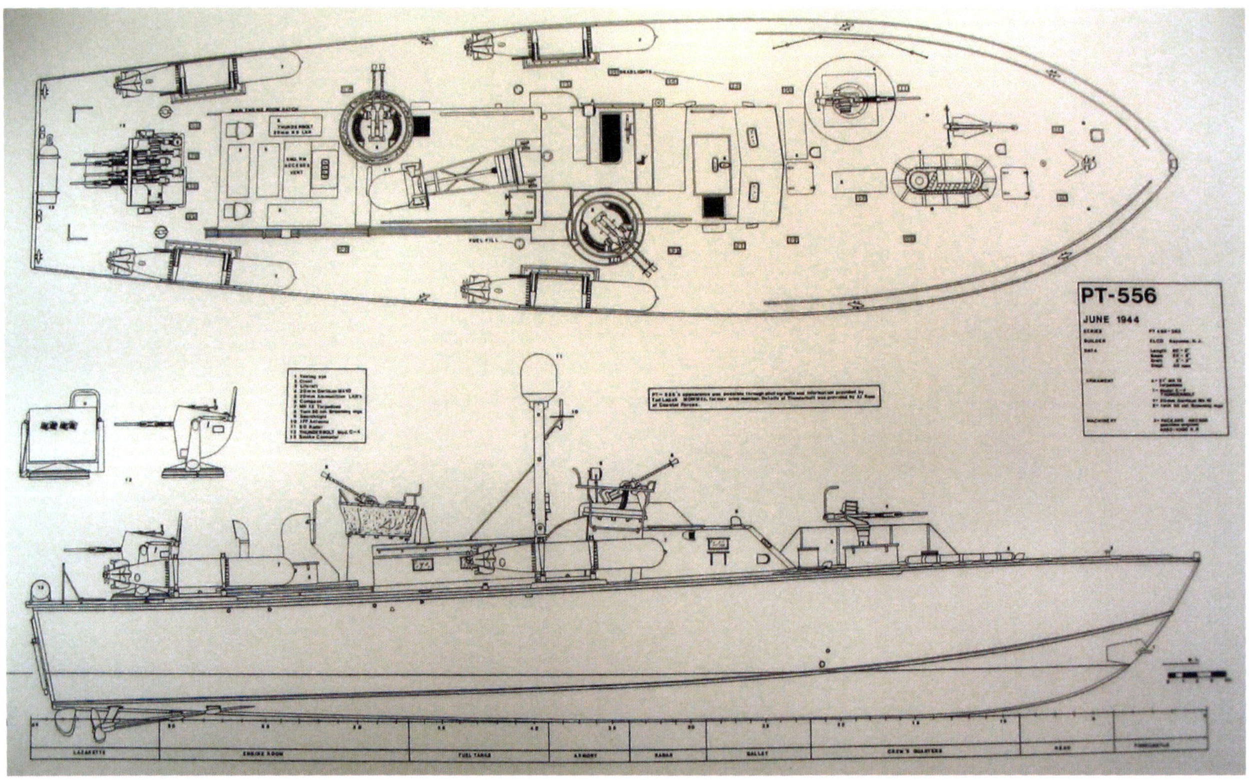

***PT-556* armed with Mark XV quad-20-mm gun mount.**
(Drawing by and used with the permission of Wayne Traxel)

The completed cockpit and charthouse section of the model. Research shows that the boats of Squadron 29 were named after characters in the popular comic strip "Lil' Abner." Although not shown here, they carried the names and an image of the character on the starboard bulkhead of the charthouse. (Steve Sobieralski photo)

The .50 caliber machine gun mounts were used from the kit with some photo-etched brass enhancement added. The "kill" markings and front port light covers are seen in reference photos of *PT-559*.

Details of the forward machine gun turret on PT-557.
(Collection of T. Garth Connelly via Frank Andruss, Sr.)

EIGHTY-FOOT ELCO PT BOAT SUPERSTRUCTURES

Seven drawings by Robert Ferrell for PT Boats, Inc. illustrate how the boats of the *PT-103* through *PT-196* series of the Elco eighty-foot boats differed from the later *PT-565* series. The differences between the two production series are readily apparent, most notably the angles of the turrets and the location of the forward starboard turret.

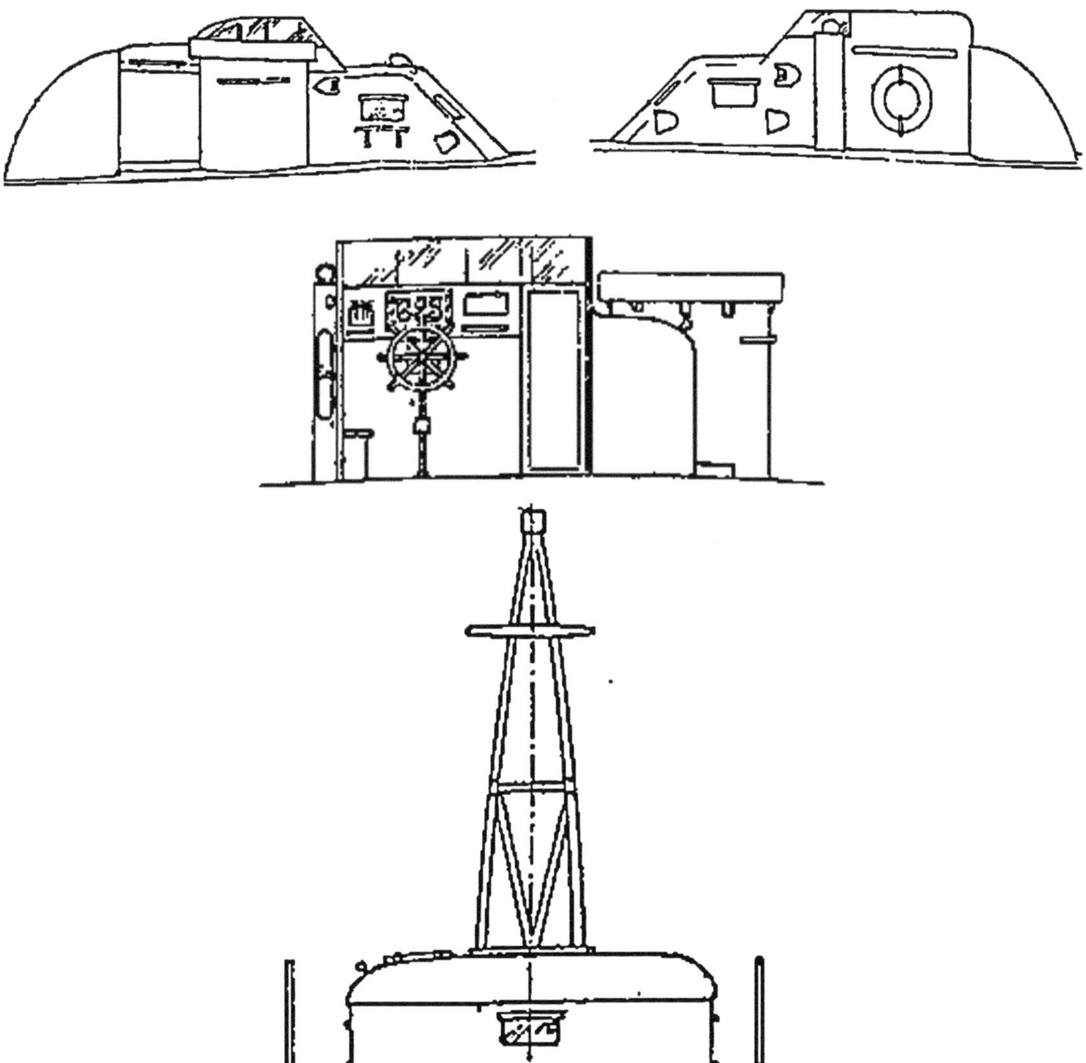

Views of the upperworks of *PT-103* Class boats.
(PT Boats Museum and Archives, Germantown, TN)

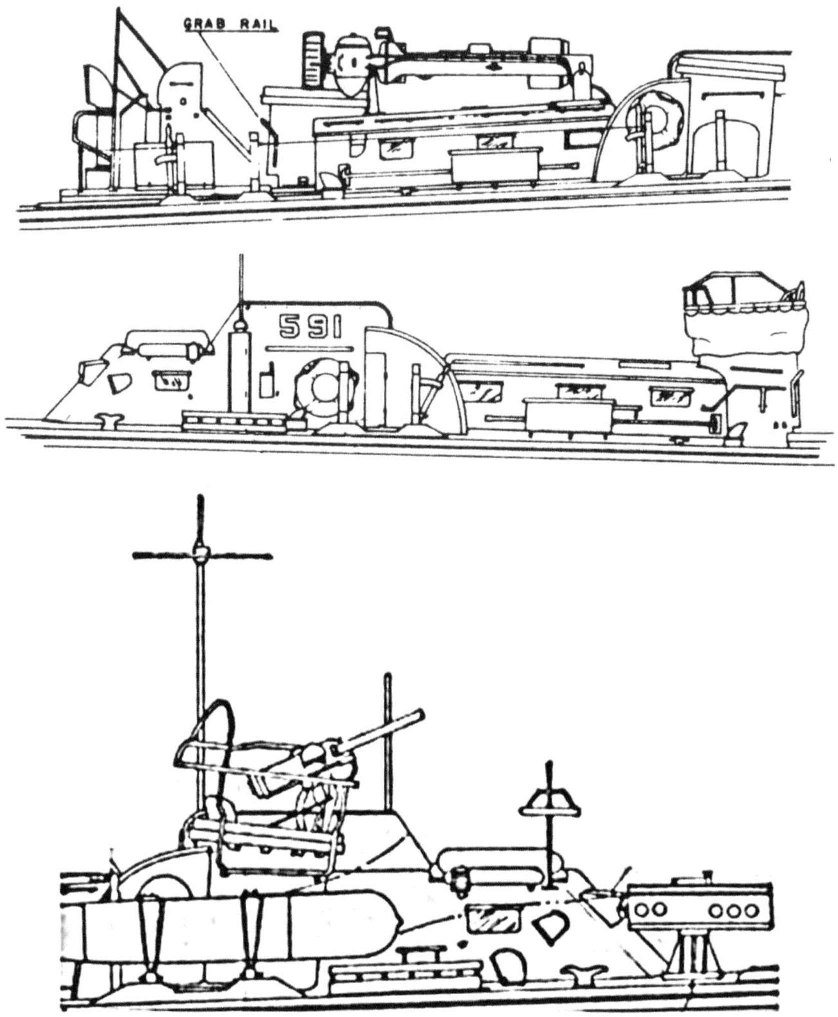

Views of how the upperworks of the *PT-565* series, which *PT-596* was a unit of. (PT Boats Museum and Archives, Germantown, TN)

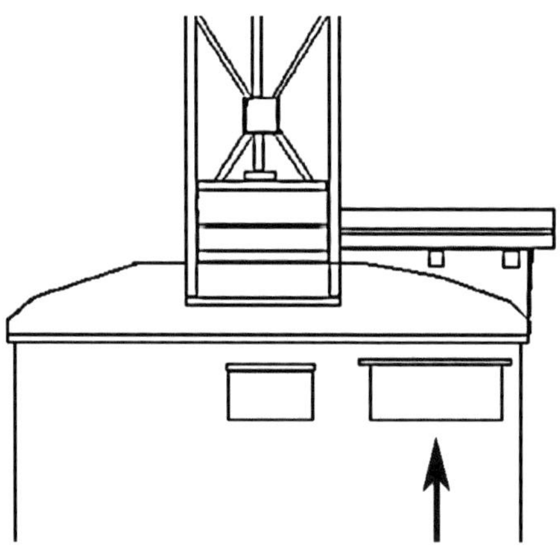

The front of the day-room cabin, looking aft to illustrate the signal-flag box location. The signal flag box is to the left of the port window in the center of the structure.
(Drawing by T. Garth Connelly, not to scale)

PT-191

Mr. Sobieralski's *PT-109* demonstrates that it is possible to backdate the Italeri kit to a first production series boat such as the *109*. For those who may wish to backdate the kit to another boat in that same production series, a boat from Squadron 12, which saw action in New Guinea and in the Philippines, *PT-191*.

PT-191 gained fame when she and *PT-190* fought off thirty to forty Japanese aircraft and shot down four of the enemy before rescue by USAAF P-47 fighters on December 27, 1943. For those who might be interested in doing this boat, modeler Alex Johnson backdated the 1:35 scale Italeri *PT-596* kit to depict *PT-191*. His modifications are described in the next section entitled *PT-191*.

***PT-191*, left, and *PT-190*, right, at Morobe River Base in October 1943. It appears that the Measure 31/5P scheme was, in fact, carried up to above deck level on the 191 as evidenced by that it can be seen on her Day-room. (Collection of T. Garth Connelly)**

This conversion involved a considerable amount of backdating by Alex Johnson to represent *PT-191*, one of the last boats produced in the original *PT-103* series. The single biggest task was repositioning the two gun turrets to represent boats prior to *PT-372*. Both turrets were angled forward slightly, and the forward turret was moved back. The tubs themselves were made using PVC pipe with a thin sheet of styrene wrapped around to create the right diameter.

Once the turrets were faired in, the modeler fabricated the rather unusual sets of depression rails found on the turrets of Squadron 12 boats. He also moved the positions of the side windows on the charthouse to their correct positions and added the running lights and extra cabin vents. Cabin handrails were trimmed and positioned in their proper places. Minor changes to the radio antenna housing, cockpit walls, and spray shields were made. A photo-etched instrument panel for the early eighty-footers was installed along with other details in the cockpit area.

Alex Johnson's completed model of *PT-191*. The model depicts the boat as she appeared on December 27, 1943, wearing the Measure 31/5P camouflage scheme. (Alex Johnson photo)

The new turret tubs were scratch-built. (Alex Johnson photo)

The original Squadron 12 boats constituted the final boats in the *PT-103* series and were the first built with a taller engine room scoop, which debuted with *PT-187*. The modeler modified the kit scoop to the correct size by reducing it to the proper width. Squadron 12 boats also carried the SO radar, another apparent first. The radar set for *PT-191* was scratch-built with plastic and cast resin parts.

The starboard turret had to be repositioned. (Alex Johnson photo)

Topside conversion work entailed scratch-building a bull-nose of a smaller size than the kit provides. The modeler made additional photo-etched deadlight frames, 20-mm ammunition boxes and the earlier style deck vents. The torpedo rack bases were modified slightly to represent the earlier style. *PT-191*'s twin .50 caliber bow mount was made with Academy guns and a scratch-built stand. Single .30 caliber guns also came from Academy on scratch-built stands. Alex Johnson made the aft Mark X mount using a spare 20-mm gun and a styrene tripod mount. The kit's smoke generator was attached to the transom along with two HR Products depth charges.

Before receiving her Bofors 40-mm gun, the *191* carried a 20-mm Oerlikon on a Mark X mount. (Alex Johnson photo)

With conversion work completed, the model was painted in a Measure 31/5P camouflage scheme. A flag, dry transfer numbers, and a scoreboard of enemy kills finished off the model. Finally, *PT-191* was "weathered" to take away the factory-fresh look often seen in models. It took a considerable amount of work, but the model builder was pleased to see that the Italeri kit could be made into an original *PT-103* series eighty-foot Elco model.

The *191*'s scoreboard can be seen along with the scratch-built SO-Type radar mast. (Alex Johnson photo)

Part of the cockpit section of *PT-191*. The man to the left of the photo is Lt. Donald Clarke, *PT-191*'s last skipper. (Collection of T. Garth Collection)

Two views of the Roll-off rack for a Mark XIII torpedo. (Collection of T. Garth Connelly)

The *191* carried her smoke generator behind two Mark VI depth charges.
(Alex Johnson photo)

At the time of the battle with the thirty to forty Japanese aircraft, the *191* had a field-mounted twin .50 caliber machine gun and a 20-mm Oerlikon gun on a Mark XIV mount on her foredeck. (Alex Johnson photo)

The completed model of *PT-191* by Alex Johnson. At the time of her famous battle a field-mounted single .30 caliber machine gun was on her day-room roof, to starboard. (Alex Johnson photo)

PT-374

Since the Italeri *PT-596* depicts a boat which never saw action, modelers might want to build a model of a boat which actually saw combat, so another option is to build the kit as a boat from a mid-war production series as it might have looked at a point late in the war.

One option can be to build the kit as *PT-374* as she would have looked at Green Island in the Solomons in 1944. For this project, Alex Johnson built a waterline model of *PT-374* for a diorama.

PT-374 was converted for use in a diorama made for Frank Andruss' Mosquito Fleet Exhibit. Alex Johnson backdated the Italeri kit for use in a scene depicting a PT boat docked at her Pacific base. Accordingly, he cut the hull at the waterline so it later sat on simulated water in the diorama. Interestingly, *PT-374* was just the third boat built with the later gun turret angles and positions. While the turrets did not need to be modified, there was still a considerable amount of conversion work to represent *PT-374*.

Alex Johnson's *PT-374* in the diorama. The layout is reminiscent of a scene from the 1963 movie *PT-109* starring Cliff Robertson. (Alex Johnson photo)

The charthouse windows, extra cabin vents, handrails, and earlier Elco running lights were repositioned. There were a number of minor changes in the bridge area, including the position of the radio antenna housing, searchlight mount, cockpit walls, and spray shields. The earlier style instrument panel was made of photo-etched brass.

Extensive modifications to the kit's cockpit and charthouse sections were needed. (Alex Johnson photo)

The modifications and enhancements made to the kit can be seen. (Alex Johnson photo)

The kit's engine room scoop was corrected to the thinner style used prior to *PT-565*. Also, a scratch-built SO radar set and the bull-nose on the bow were added. The front hatch was ultimately left open and a figure placed in the opening to represent a crew member entering the crew's quarters.

Armament included the four torpedo racks, which were modified slightly to represent the earlier style bases. In the diorama, men are loading a torpedo in the aft starboard rack. On the bow, the 37-mm gun was converted from the Italeri kit's M9 mount to the earlier M4 version. Sheet plastic was used to make the short base, and photo-etched loops were used to represent the shape of the earlier ammunition magazine. *PT-374*'s 60-mm mortar came from an Academy set of US infantry weapons.

The scratch-built mount for the 37-mm M4 gun on *PT-374* can be seen here. The Academy 1:35 scale 60-mm mortar was also used. (Alex Johnson photo)

The depth charges seen on this model came from HR Products. (Alex Johnson photo)

PT-374's SO radar set was built from scratch with resin and plastic parts. The Mark XVII mounts in the turrets were super-detailed with photo-etched ammunition belts by Eduard. Two HR Products depth charges were placed amidships between the sets of torpedo racks. The field-installed set of twin .30 caliber machine guns on the aft starboard torpedo rack was built with Academy guns on a mounting made of scrap plastic.

On the fantail, the boat's 40-mm gun was installed with very slight modification. A brace was added to secure the gun's barrel when not in use. Two 40-mm ammunition stowage boxes were built from sheet plastic. One was left open to show the box's interior for the diorama.

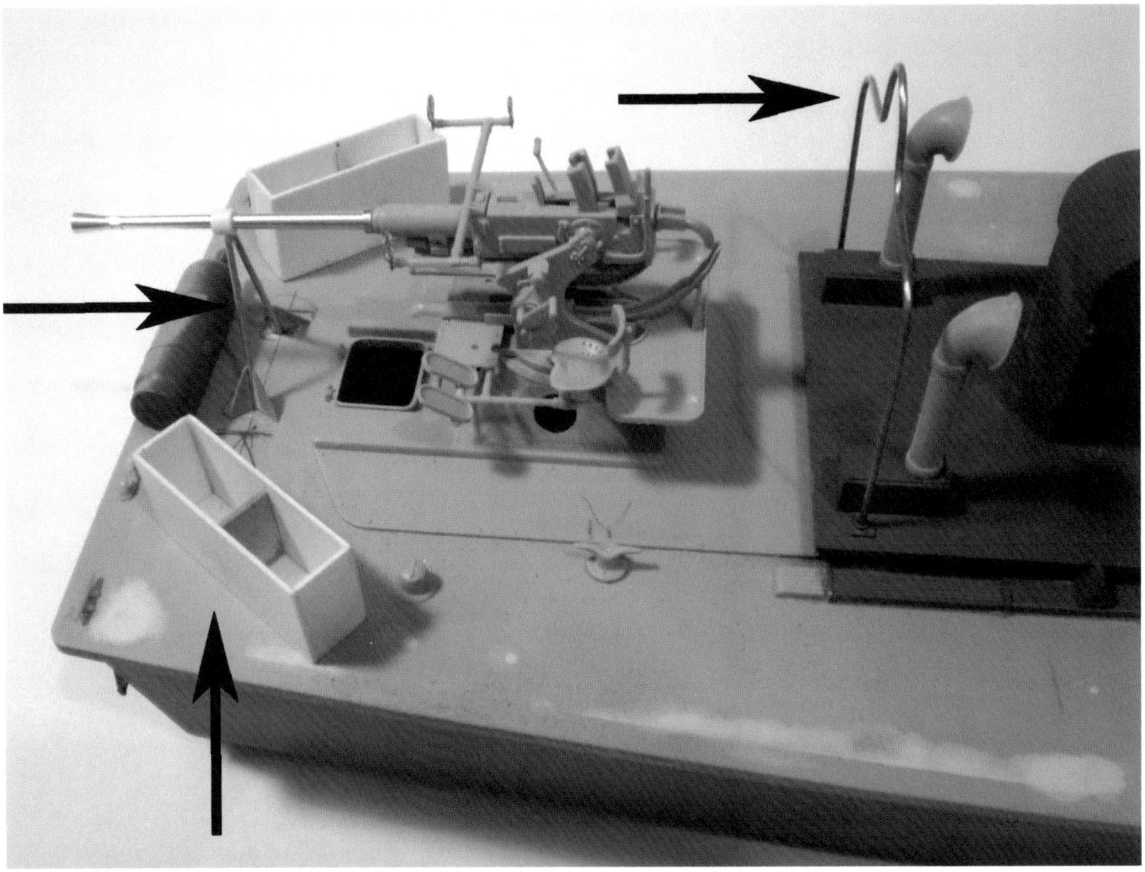

The two ready-service lockers were scratch-built along with the depression-rail for the Bofors 40-mm gun and the bracing for the barrel. (Alex Johnson photo)

Following the application of Measure 31/5P camouflage paint scheme on the model, final details included application of the flag and boat numbers. Minor equipment was added including a toolbox by the engine room hatch, a set of binoculars in the cockpit, and rope coils on deck. Fellow modeler Bridge Carney crafted a dozen life vests, which dressed up the superstructure. Made of tea bags, the vests were amazingly lifelike. The modeler also gives great credit to the figures painted by Frank Ryczek and Al LeFleche. Lastly, the entire boat was weathered to give it that added touch of realism before being placed into the diorama.

***PT-374* wears Measure 31/5P as her camouflage scheme. (Alex Johnson photo)**

Al LeFleche and Frank Ryczek painted the figures in the diorama. Frank Ryczek built the buildings and vehicles. The life jackets were made by Bridge Carney. (Alex Johnson photo)

There is a sense of activity in this diorama. (Alex Johnson photo)

MARK XII MOUNTS

In the first production series boats, Mark XII 20-mm mounts were fitted on some boats. On those boats which were fitted with them, the mounts were installed in the forward turrets in lieu of the twin .50 caliber Browning M2 machine guns.

The boats with these mounts were *PT-115*, *PT-154* through *PT-157*, *PT-160*, *PT-162*, *PT-168,* and *PT-187* through *PT-195*. These boats all carried the Mark XII mounts in their forward turrets. There is photographic evidence to indicate that *PT-156* and *PT-195* also had this mount in their aft turrets.

***PT-195* shown with the Mark XII mounts for the 20-mm guns in both turrets. (Will Day photo)**

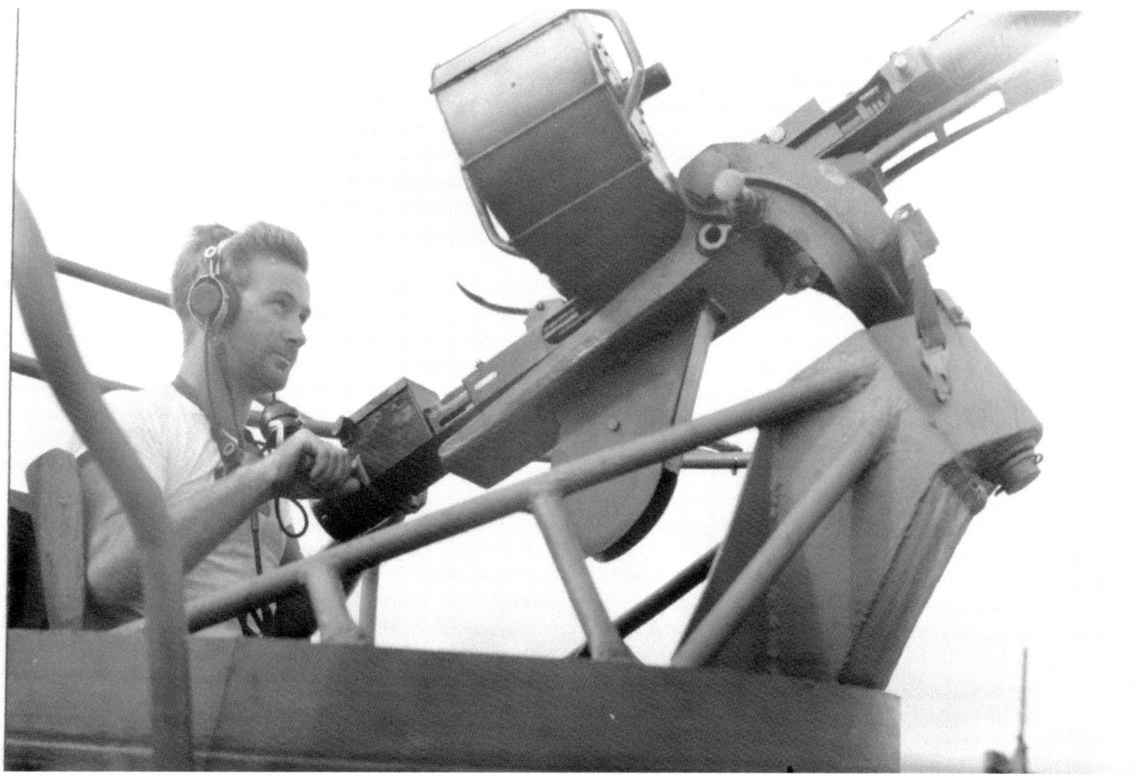

Mark XII mount for an Oerlikon 20-mm gun. (B.G. Marshall)

Wayne Traxel has indicated that *PT-191* did experiment with this mount while she was fitting out in the United States but reverted to the machine guns before she was shipped out. Backdating the Italeri kit to depict one of these boats might prove to be an interesting project to undertake.

The Mark XII mount in the forward turret of *PT-157*.
(T. Garth Connelly via Wayne Traxel)

BAZOOKA ARMED PT BOATS

The sextuple bazooka was an interesting type of weapon when fitted to a PT boat. Squadron 25, which was in the Panama Canal Zone before being sent into action, decided to experiment with this weapon. They requested to be allowed to mount a sextuple bazooka on a tripod mount, which was similar to those fitted to Elco seventy-seven foot *PT-20* Class boats. The Bureau of Ships approved this request in 1943. The mount was constructed and tested by a Squadron 25 boat in the Panama Canal Zone. The bazooka was mounted onto the roof of the day-room, just behind the radar mast. Testing showed that it would not be an effective weapon for PT boats.

The photos below were supplied by Tracy White.

The sextuple bazooka mount fitted to the Day-room roof behind the radar mount of an unknown Elco eighty-footer. The experimental fitting of this weapon was not a success and it was not used on PT boats. (Collection of T. Garth Connelly)

CAMOUFLAGE

Boats that were heading to the Pacific carried a camouflage scheme named Tropical Green System. This scheme consisted of overall Navy Green (5-NG) or Tropical Green. Another scheme was Measure 13, which consisted of overall Haze Gray (5-H) with Deck Blue (20-B) decks. The measure designated as Measure 14 was similar to Measure 13 but consisted of Ocean Gray with Deck Blue (20-B) decks.

***PT-196* in a non-standard scheme before being shipped to New Guinea with Squadron 12. The boat was repainted in Measure 31/5P before leaving the US. (Collection of T. Garth Connelly)**

The boats of Squadron 29 were originally painted in Measure 31/5P as they were originally assigned to the Pacific. Measure 31/5P had a base color of Ocean Green (5-OG) and irregular areas of Navy Green (5-N) with either hard or soft edges. Sometimes the irregular areas were carried over to the decks, but this proved too time-consuming to maintain at the front lines, so the decks were painted in Deck Green (20-G).

Before these boats were sent to the Pacific, however, the Navy reassigned the Squadron to the Mediterranean Theater of Operations. They were repainted in the Thayer Blue System, which consisted of an overall scheme of Thayer Blue (5-B) and Deck Blue (20-B). It was known as Measure 16.

***PT-557* at speed in an overall gray scheme while at Melville. (Collection of T. Garth Connelly)**

The boats, which saw action in that theater, had a three-foot triangle painted on their foredecks, which went from the forepeak back three feet. These triangles were painted yellow for air recognition purposes. Additionally, there is now anecdotal evidence that boats also had a red colored stripe/panel at the stern.

However, there is photographic evidence to support a theory that not all of the boats had this red stripe/panel. Therefore, modelers should do copious research to determine if the boat they are building had the stripe.

If modelers select to do a *PT-103* Class Elco eighty-footer, they could opt to do a boat of Squadron 12. This Squadron was based in New Guinea in late 1943 through late 1944. These boats were fitted with additional weapons and were painted in Measure 31/5P.

Measure 31/5P consisted of MTB Green, Navy Green (5-NG) and Deck Green (20-G).

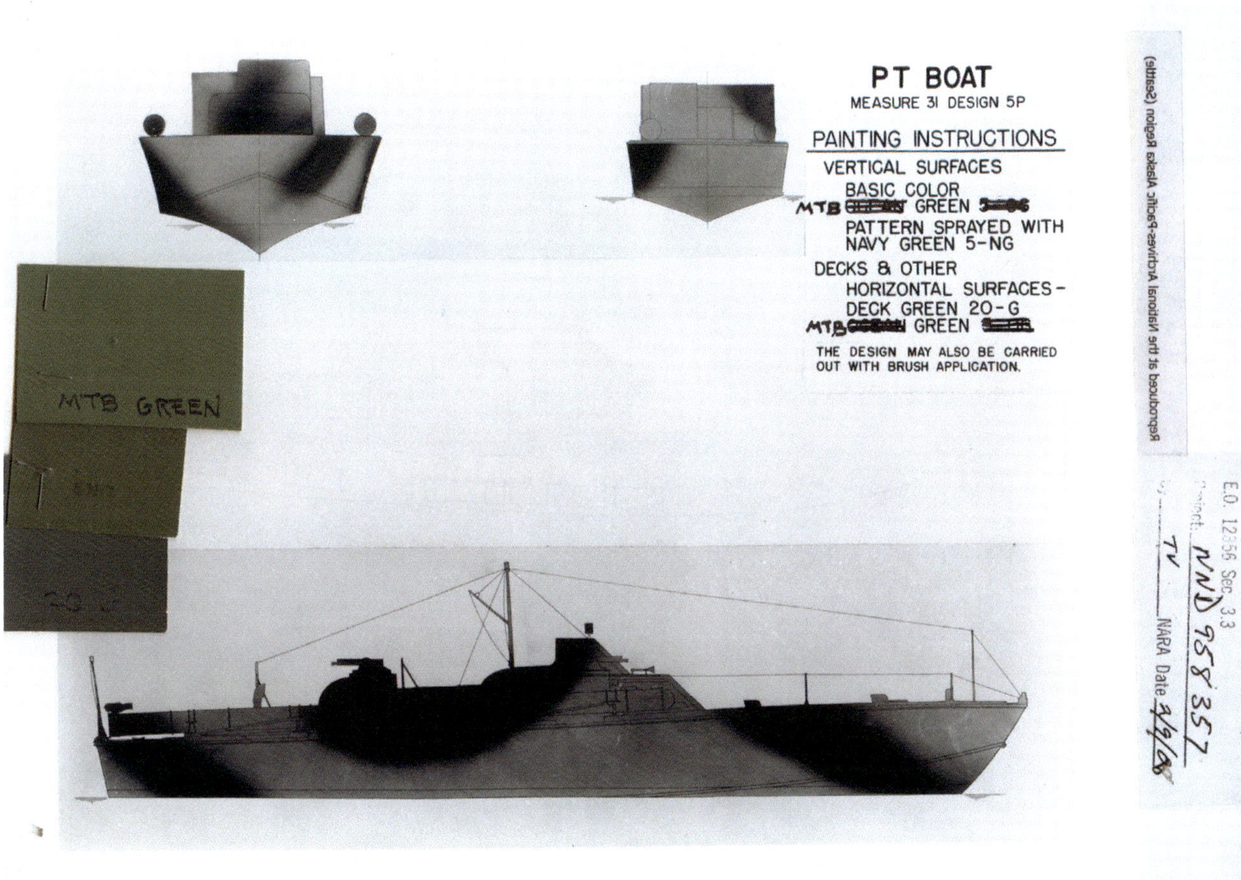

Painting instructions for USN Camouflage Measure 31/5P. (Author's Collection via Tracy White)

Measure 31/20L camouflage scheme consisted of the following colors:

- Green 2
- Green 3
- Dull Black
- Deck Green
- Copperoyd (a substitute can be Norfolk 65-A Anti-fouling Red, from White Ensign Models' Colourcoats Line)

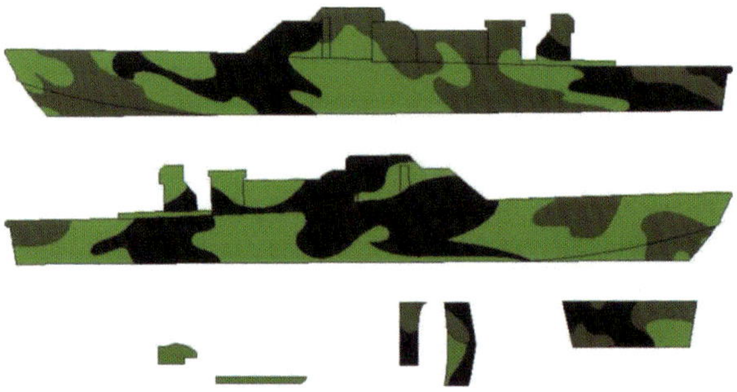

The colors of Measure 31/20L. Drawing used with the permission of Kenny Loup of GatorMasks (www.gatormasks.com)

Modelers should note that Norfolk 65-A Anti-fouling Red paint was never used on wooden hulls. The bottom anti-fouling paint, applied to the bottom of the wooden hull minor combatants, was known as Copperoyd.

Other modelers might choose to backdate the kit to depict any of the boats of Squadron 10 (*PT-163* to *PT-174*). These were units of the first production series of the *PT-103* Class. Originally, the boats of this squadron were painted in a camouflage scheme known as the Adaptor Scheme, commonly referred to as the Zebra Scheme.

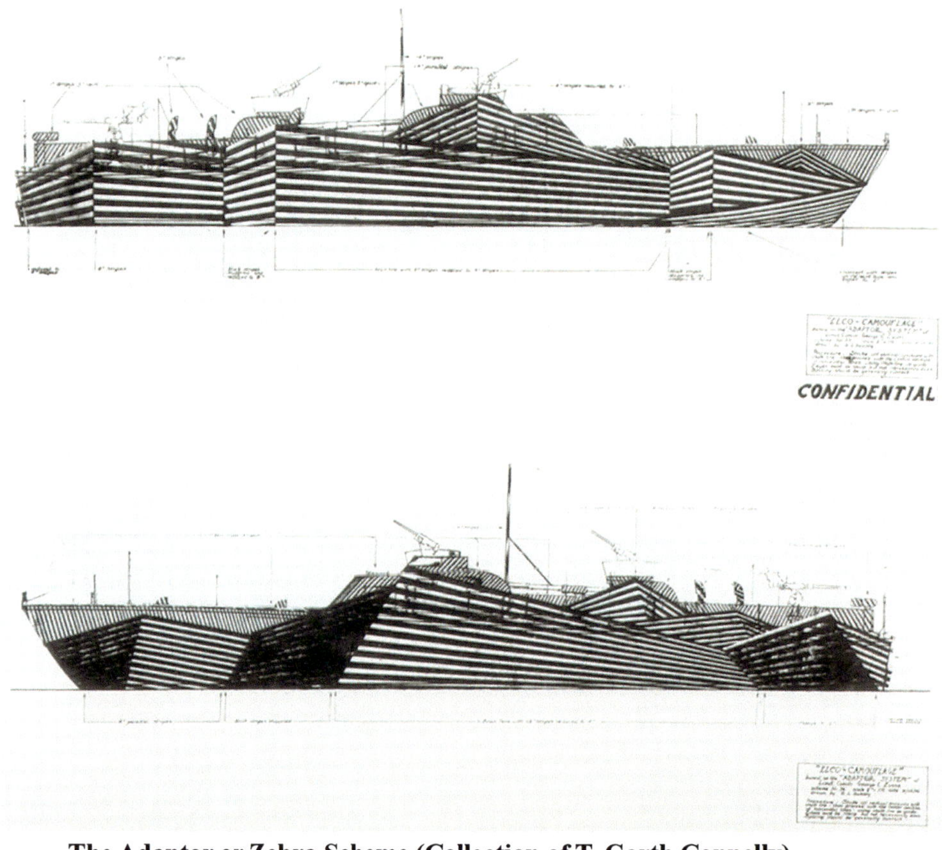

The Adaptor or Zebra Scheme (Collection of T. Garth Connelly)

***PT-174* in a variation of the Adaptor Scheme. No two boats had the same application.**
(Collection of T. Garth Connelly)

An unknown Squadron 10 boat underway in a variation of the Adaptor Scheme.
(Collection of T. Garth Connelly)

There were two versions of the Adaptor Scheme.

The White version consisted of:

White
Black
Haze Gray (5-H)
Deck Blue (20-B)

The Green version consisted of:

Haze Green (5-HG)
Black
Deck Green (20-G)

A man who had first-hand knowledge that the boats of Squadron 10 were painted in the Green version of the Adaptor Scheme wrote a handwritten document available through PT Boats, Inc.

PT-169 **showing another variation of the Adaptor Scheme. Note that she is not fitted with any of her torpedo tubes. (Collection of T. Garth Connelly)**

PT-170 **in another variation of the scheme. Note that the canvas covering for the turret is dyed to match. (Collection of T. Garth Connelly)**

PT-171 **in still another variation of the Adaptor Scheme. Some of the panels in all of the variations were painted in a counter-shading bluish-gray.**

The counter-shading mentioned in the caption for *PT-171* was achieved with light blue-grayish counter shading. The horizontal surfaces called for Deck Blue (20-B).

RADAR

By the midpoint of World War II, USN PT boats carried the SO-Type radar. Toward the end of the war, the SO-3 Type radar was fitted. The SO Type had the familiar dome as seen in photos of mid to late war boats. The SO-3 Type was fitted to late war (1945) boats. This radar had the "football-shaped" dome.

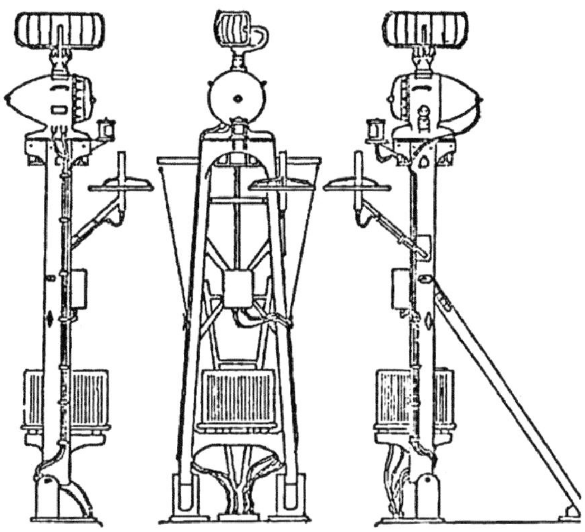

The SO-3 Type radar as carried on the later series of PT boats. (PT Boats Museum and Archives, Germantown, TN)

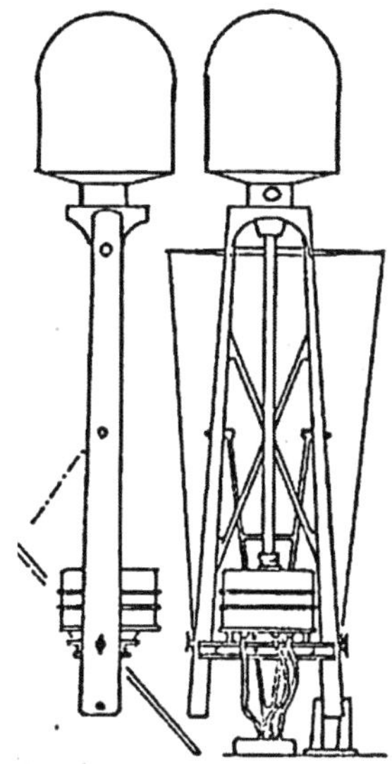

SO-Type radar as seen on all USN PT boats from mid-1943 to late 1944. (PT Boats Museum and Archives, Germantown, TN)

By late 1942, experimental radar was fitted to some boats to give them a tactical advantage in the field and to see if the boats could utilize radar. That radar was called the SCR-517A. This unit was built by Westinghouse, but Bell Labs was also involved with development. It was an Army S Band Surface Radar that was used on PT boats. A total of 225 were built. It was developed from the Black Widow night-fighter's radar.

According to B.G. Marshall, the boats having the SCR-517A installed were *127*, *128*, *131*, *132*, *135*, *136*, *142*, *143*, *145*, *149*, *151*, and *152*. Modeler and draughtsman Wayne Traxel provided additional data that another boat, *PT-164,* also carried the SCR-517A. *PT-164* was one of the boats painted in the Adaptor scheme.

It has also been learned from Alex Johnson that two boats of Squadron 9, *PT-152* and *PT-155,* had their units installed at the Elco factory. Ted Walther added that *PT-167* carried the SCR-517 radar as did the *168* and *169*. The *174* carried a SO-A radar. Mr. Walther believes that *PT-166* carried the SCR-517 radar, but there is no photographic evidence of this.

Front view of the SCR-517A radar on *PT-149*. It is readily apparent as to why it was called the "Beehive" radar. (Collection of T. Garth Connelly)

The SCR-517A "Beehive" radar seen on *PT-149* as fitted onto the original style mast using additional bracing. (Photo courtesy of Will Day)

The SO-Type Radar started to appear on PTs from mid-1943 onwards. The late war boats (late 1944 to 1945) started to come from the factories complete with the SO-3 Type radar installed.

SO-Type radar fitted to an Elco eighty-footer late in the war. (Collection of T. Garth Connelly)

***PT-167*'s SO-Type radar. Note how it is devoid of "yard-arms." (Collection of T. Garth Connelly)**

***PT-191's* SO-Type radar with stubby "yard-arms." (Collection of T. Garth Connelly)**

Scratch-built SCR-517A radar by Alex Johnson. Note how the bracing for the unit was made. (Alex Johnson photos)

The SO-Type radar fitted to *PT-545.* Note the lack of a B.N. antenna and "yard-arms."
(Collection of T. Garth Connelly)

The SO-Type radar fitted to *PT-156* did not have the B.N. antenna on the mast. (Collection of T. Garth Connelly)

***PT-328* as refitted with SO-3 radar. (Collection of T. Garth Connelly)**

AVAILABLE AFTERMARKET ACCESSORIES FOR THE 1:35 SCALE ITALERI *PT-596* KIT

For those modelers who wish to take their models from this kit to the next level and super-detail their models, there are a number of items to assist them with doing this.

Aber .30 and .50 caliber machine-gun ammunition belts (stock #35A47)

Archer Dry Transfers, decal sheet (#AR35238 and #AR35239)

Master Model, brass barrels

Italeri figures (kit/part #5606), 10 figures

G-Factor Elco PT Boat Propellers (stock #13501)

Lion Roar detail set for the *PT-596* kit. The set contains the following: 9 X frets P-E parts, 1 X P-E name plate, 3 X brass propellers, 16 X metal rocket-launcher tubes, 1 X instrument panel film, 6 X various metal gun barrels, Approx 49 X various gun cartridges (Lion Roar LA 35003)

Eduard detail set for the kit (ED53025)

Eduard ammunition belt, colored (TP509)

Wooden pier from Royal Models (RYMRM531)

Mobile crane from Royal Models (RYMRM532)

USN sailors loading torpedo from Royal Models (RYMRM538)

USN sailor washing deck from Royal Models (RYMRM537)

Lion Roar barrel for 40-mm gun (LA 3531)

Lion Roar barrel for 20-mm gun (LA 3530)

Generator from Royal Models (RYMRM529)

Elco 80-footer accessories (life jackets, helmets) and dock accessories from Royal Models (RYMRM535)

USN PT boat officer and sailor from Royal Models (RYMRM536)

Detail PE set for the *PT-596* kit from Royal Models (RYMRM561)

VLS Warriors Scale Models, two figures (skipper and enlisted man helmsman) (WA35628)

VLS Warriors Scale Models, 37-mm gunner (WA35630)

VLS Warriors Scale Models, .50 caliber machine gun gunner (WA35637)

VLS Warriors Scale Models, .50 caliber machine gun gunner (WA35636)

U-Models, depth charges (UM063)

REFERENCE MATERIAL

The following available books are excellent sources for reference on PT boats. This is not a full or comprehensive listing but serves as a suitable resource. Additionally, the exhibits of PT boats at Battleship Cove in Fall River, Massachusetts, offer excellent sources for reference.

American PT Boats in World War II by Victor Chun Schiffer Publishing, Military/Aviation History, ISBN 0764302566

Allied Coastal Forces Of World War II: Volume II Vosper MTB's & U.S. Elco's by John Lambert & Al Ross, Conway Maritime Press/United States Naval Institute Press, ISBN 0851776027

PT Boats in Action: Warships Number 7 by T. Garth Connelly, Squadron/Signal Publications, 1994, ISBN 0897473124

U.S. Small Combatants, including PT Boats, Sub Chasers and the Brown Water Navy: An Illustrated Design History by Norman Friedman, with ship plans by A.D. Baker, III, with Alan Raven and Al Ross, United States Naval Institute Press, 1987, ISBN: 0870217135

United States PT Boats of World War II in Action by Frank D. Johnson, Blandford Press, UK, 1980, ISBN 9780713710250

At Close Quarters PT Boats in the United States Navy by Captain Robert J. Bulkley, Jr., USNR (Retired), Naval History Division, Washington, 1962 (Government Printing Office)

U.S. Patrol Torpedo Boats in World War II, 1939-1945 by T. Garth Connelly, Nimble Books, LLC, 2010, ISBN 978160880133

Building the PT Boats by Frank J. Andruss, Sr., Nimble Books, LLC, 2009, ISBN 9781934840856

There are also a number of websites which could prove helpful to modelers:

www.ptboatworld.com
www.ptboats.org
www.pt103.com
www.pt171.org
www.shipcamouflage.com

ABOUT THE AUTHOR

T. Garth Connelly is a naval historian who specializes in minor combatants. He has written five books since 1994, *PT Boats In Action, Vosper MTBs In Action*, *U.S. 110' Subchasers In Action*, *U.S. Patrol Torpedo Boats In World War II 1939-1945*, and he co-authored *Schnellboot In Action.* He also wrote *The Regia Marina's MAS Boats, 1915-1945* with the assistance of the former Naval Attache at the Italian Embassy and the Italian Navy. He has also written several articles that have appeared in *Model Ship Builder*, *Ships In Scale*, *Model Ship Journal* and the International Plastic Modelers' Society's *Modelers Journal* since 1996. He was awarded the 2000 Clean Sweep Award by the Veterans Association known as PT BOATS, INCORPORATED because of his continuing effort to promote the history of PT boats in World War II.

CPSIA information can be obtained
at www.ICGtesting.com
Printed in the USA
255246LV00009B

* 9 7 8 0 9 8 4 1 2 6 7 5 0 *